Soil erosion, an agricultural disaster, Locust Grove, Sherman County, Oregon (Frank M. Roadman, USDA–Soil Conservation Service photograph).

Springer Series on Environmental Management

Robert S. DeSanto, Series Editor

Harold D. Foster

Disaster Planning

The Preservation of Life and Property

With 48 Figures

Springer-Verlag
New York Heidelberg Berlin

Harold D. Foster
University of Victoria
P.O. Box 1700
Victoria, British Columbia
Canada V8W 2Y2

According to German legend the Lorelei lured sailors to their deaths on the Rhine rocks. This book is dedicated to a Lorelei who, despite her name, has helped me avoid many a disaster.

Library of Congress Cataloging in Publication Data
Foster, Harold D
Disaster planning.

(Springer series on environmental management)
Bibliography: p.
Includes index.
1. Disaster relief—Planning. I. Title.
II. Series.
HV553.F64 363.3'48 80-18910

Printed in the United States of America.

9 8 7 6 5 4 3 2 1

ISBN 0-387-90498-0 Springer-Verlag New York
ISBN 3-540-90498-0 Springer-Verlag Berlin Heidelberg

Series Preface

This series is dedicated to serving the growing community of scholars and practitioners concerned with the principles and applications of environmental management. Each volume will be a thorough treatment of a specific topic of importance for proper management practices. A fundamental objective of these books is to help the reader discern and implement man's stewardship of our environment and the world's renewable resources. For we must strive to understand the relationship between man and nature, act to bring harmony to it and nurture an environment that is both stable and productive.

These objectives have often eluded us because the pursuit of other individual and societal goals has diverted us from a course of living in balance with the environment. At times, therefore, the environmental manager may have to exert restrictive control, which is usually best applied to man, not nature. Attempts to alter or harness nature have often failed or backfired, as exemplified by the results of imprudent use of herbicides, fertilizers, water and other agents.

Each book in this series will shed light on the fundamental and applied aspects of environmental management. It is hoped that each will help solve a practical and serious environmental problem.

Robert S. DeSanto
East Lyme, Connecticut

Preface

The daily newspapers record a never-ending series of disasters. From epidemics to invasions, each headline is accompanied by graphic descriptions of death, suffering, and destruction. Nevertheless, many communities tend to ignore the risks to life and property posed by the ever-increasing hazard spectrum. Others make token efforts at preparedness, accepting the possibility of local disaster and designing a plan to deal with its aftermath. A few, more enlightened municipalities have begun to recognize that every decision has an impact on risk, and therefore on the probability of disaster. This book is written to encourage more local authorities, institutions, and organizations to accept their responsibility to increase safety through such comprehensive risk management.

There will be those no doubt who, after reading this volume, will exclaim, "the price of safety is too high." There are few goals that are worth satisfying at the expense of community destruction or dramatic increases in deaths and illnesses. In response to such criticism it should be pointed out that in urban communities the staff, equipment, and information already may be available to implement many of the stages of the disaster mitigation model presented in this volume. What is often needed is a redefinition of roles, change in emphasis, and a commitment to achieve safety goals. In many cases procedures and practices may be made less hazardous by changes that are not necessarily more expensive. Similarly, if through the implementation of a safety plan, death, illness, and destruction are reduced, other community programs will become less financially demanding. Where additional assistance is required to meet safety goals, this may often be provided at minimal cost by senior levels of government, or by college and university departments. For example, postdisaster plans, designed to speed recovery if destruction should occur, can be prepared as training exercises for urban geographers, planners, and architects.

The situation is more complicated in rural areas, where resources and expertise are generally limited. Since safety is everyone's concern it should also be each individual's responsibility. For this reason, local officials might actively seek assistance from various community groups on a volunteer basis.

Some of the concepts in this volume are my own and I take full responsibility for them. The large majority, however, have been gleaned from the published works of the small army of dedicated individuals who attempt to reduce the suffering of others by an analysis of the causes and consequences of disaster. To all such colleagues I owe a considerable intellectual debt. Their names are listed in the bibliographies following each chapter.

I am also pleased to acknowledge the encouragement of Dr. Robert S. DeSanto, who first suggested I write such a book. Mrs. E. Lowther has been of great assistance in her continuing role as the world's best typist. Lorelei, my wife, has been invaluable as both proofreader and constructive critic. I should also like to acknowledge the drafting and graphic skills of Mr. Ian Norie, Mr. Ole Heggen, and Mr. Ken Quan, which have added greatly to the quality of the final product.

Harold D. Foster
University of Victoria
Victoria, B.C.

Contents

1
Introduction

What plagues and what portents, what mutiny
What raging of the sea, shaking of the earth,
Commotion in the winds, frights, changes, horrors,
Divert and crack, rend and deracinate
The unity and married calm of states

Ulysses in *Troilus and Cressida*
Act 1, Scene iii
William Shakespeare (1564–1616)

It is impossible to avoid all risk. As a result, societies have evolved in a manner that allows them to operate within specific levels of tolerance for natural and man-made events. Typically those boundaries to what can be successfully accommodated are defined either by law or by common practice. Often regulations, such as public health and building codes, identify the maximum event that must be guarded against. The resulting level of socially acceptable safety reflects such factors as past experience, needs, wants, and wealth. Regardless of the emphasis placed on the preservation of life and property, there will always remain the potential for the occurrence of extreme events, capable of overcoming the capacity of society to cope without dramatic changes in its normal operation. Such threats are known as hazards, and their impact on society as disaster. While hazards cannot be eliminated, the limits of tolerance of every society to their effects can be increased and hence the potential for disaster reduced. Mankind faces a wide and expanding spectrum of hazards capable of causing death and destruction. For the majority of the world's population, living in the rural areas of the Developing World, the hazards that threaten their livelihood are still the ancient ones which commonly afflict agriculture and settlement. These are predominantly of natural origin (Harriss, Hohenemser, and Kates, 1978). For example, the losses from geophysical hazards, such as tropical cyclones, earthquakes, floods, and droughts each year in the Developing World cause an average of 250,000 deaths and $15 billion in damage, prevention, and mitigation costs (Burton, Kates, and White, 1978). This is the equivalent of 2–3% of the gross national product of those countries affected by such disaster agents.

Losses from crop disease, vermin, and pests are problems of even greater proportions. In addition, infectious disease, though declining in impact, still accounts for between 10 and 20% of human mortality. This is especially prevalent among the very young (World Health Statistics Annual, 1976).

Natural hazards are a relatively less significant issue in industrialized nations. In the United States, for example, geophysical hazards are responsible for less than 1000 annual fatalities, while resulting property damage, together with costs of prevention and mitigation, are the equivalent of 1% of the gross national product. Crop diseases and the ravages of pests and vermin are controlled by a wide variety of insecticides, pesticides, and other techniques (Harriss, Hohenemser, and Kates, 1978). Many infectious diseases have been eliminated and together they account for less than 5% of mortality in the Developed World. Despite this progress toward safety, the place of natural threats is being taken by hazards which are increasingly rooted in the use and misuse of technology. Many of these man-made problems now have an impact comparable to or greater than that of the natural hazards they have superseded. The United States, for example, currently spends $40.6 billion per year, or approximately 2.1% of its gross national product, in attempting to reduce air, land, and water pollution and its damaging effects. The annual losses caused by automobile accidents alone is placed at $37 billion or 1.9% of the gross national product (Faigin, 1976). Harriss, Hohenemser, and Kates (1978) have estimated that between 20 to 30% of all male deaths and 10 to 20% of all female deaths in the United States stem directly or indirectly from technological hazards. In addition to the immense suffering involved, these represent an expense in lost productivity and medical costs of between 2.5 and 3.7% of the gross national product. It has been estimated that overall, expenditures and losses due to technological hazards in the United States may be as high as $200 to $300 billion, a figure that represents between 10 and 15% of the gross national product (Tuller, 1978).

Despite ever-increasing knowledge and technological sophistication, losses from both natural and man-made hazards continue to rise at least as fast as the increase in global wealth and population. It is the author's belief that this paradox is not inevitable and that losses can be contained. This task is not simple and requires the far more rigorous and widespread application of the techniques of disaster prevention, prediction, and mitigation outlined in this volume.

Unfortunately, all too often decision makers respond adequately to threat only after disaster has occurred. This is inadequate and the right to safety must be guarded with at least as much vigilance as other basic human freedoms. Indeed there are no potential infringements upon individual rights so permanent as the death, injury, or debilitating destruction which so often results from unnecessary communal neglect of natural and man-made hazards. Analogies between personal and societal disregard of risk are obvious. It is widely appreciated that the overweight, alcoholic smoker who shuns exercise mars his quality of life and becomes a prime candidate for coronary thrombosis or other fatal diseases or accidents. Similarly, the community that develops without carefully weighing the impact of its growth on risk follows a comparable path. Poorly located or controlled land use and badly designed buildings and transportation systems inevitably lead to a decline in the quality of life and an increase in the potential for major disaster. Day-to-day losses are promoted by unnecessary traffic and industrial accidents, illnesses stemming from pollution of workplace or environment, and the frequent but unaccommodated impacts of small scale natural events. Such societies also increase their potential for large natural and man-made disasters, which when they strike cause death and destruction that could have been avoided. Recovery is also retarded by an absence of preimpact planning and preparation.

Communities, like individuals, may often work toward their own destruction through neglect, ignorance, or a deliberate emphasis on fulfilling superficially advantageous short-term goals. Incrementally, in doing so, they magnify risk and eventually suffer the disasters they deserve. Consider, for example, the case of seventeenth century London. Early on the morning of 2 September 1666, a fire started at Farryner's bakery shop on Pudding Lane. The houses in this area were typical of the period, closely built, wooden shacks, full of brush and faggot wood, their projecting stories almost blocking the daylight on the ground below. The lane itself, so typical of others in the city, was narrow, filthy, and acted as an open sewer carrying human and animal wastes. Despite numerous unenforced regulations, the common people had jerry-built their homes and shops on every available location and in every conceivable manner. This lack of appreciation of risk had already led to several outbreaks of the plague, one of which had killed over 75,000 Londoners the previous year. Fires had also been commonplace. The summer of 1666 had been very dry and on the Sunday morning of 2 September there was a brisk easterly wind. In consequence, the fire in the bakery spread rapidly, springing from thatched roof to thatched roof. There was no adequate fire brigade to stop its progress and the people fled in panic. It lasted for five days, consuming four-fifths of the city, some 13,000 houses and 90 churches in all (Ferguson, 1975).

Fortunately, communities, like individuals, can reform. Changes can be made to infrastructure and function which together improve the quality of life and reduce the potential for future disaster. The Great Fire of London taught the citizens of the medieval capital that the problem of mounting risk could not simply be ignored. By the end of March 1667 the *London Building Act* had been passed by Parliament. Among other things this provided that all building should have exterior finishes of brick or stone, and be constructed on definite surveyed frontage lines. The number of stories was also controlled as was the thickness of exterior and party walls. London city authorities were required to broaden many streets and passages could not be less than 14 ft wide. The mayor was also given authority to levy taxes to meet the cost of this street widening program and sewer and drain construction. In summary, the destruction of medieval London, made virtually inevitable by an almost total lack of disaster mitigation practices, gave rise to the concept of building codes and set their future pattern for the rest of England and for North America (Ferguson, 1975).

The Great Fire of London illustrated a truism. Safety is not only a right, it is also a responsibility. The most fundamental deficiency of the contemporary approach to disaster planning is its separation from day to day decision making. Since disasters occur when the level of social tolerance to natural and man-made events is overcome, it follows that the most logical way to avoid such losses is to expand tolerance. There can be no clear-cut boundary between routine town and regional planning and disaster mitigation. Every land use decision, and indeed every social or economic policy, carries with it implications for risk. In consequence, each decision increases or decreases the potential for future disaster. It is this intimate relationship which is so frequently overlooked and, as a result, so often leads to catastrophe. If losses from natural and man-made hazards are to be reduced, then risk must be given greater cognizance by every level of government at every stage of decision making. This can perhaps best be achieved by the development of comprehensive plans which include safety as one of their major goals. In this manner all significant policies would be routinely assessed to ensure that they did not seriously impair society's ability to respond to adverse events or increase the probability of disaster taking place. The remainder of this volume is devoted to a description of interrelated strategies which can be applied to assist in such a holistic approach to disaster prevention.

References

Burton, I., R. W. Kates, and G. F. White. 1978. *The Environment as Hazard.* Oxford University Press, New York, 240 pp.

Faigin, B. M. 1976. *Social Costs of Motor Vehicle Accidents.* Report DOT-HS 802119, U.S. Department of Transportation, Washington, D.C.

Ferguson, R. S. 1975. Building codes—Yesterday and today. *Habitat,* **18(6)**:2–11.

Harriss, R. C., C. Hohenemser, and R. W. Kates. 1978. Our hazardous environment. *Environment,* **20(7)**:6–41.

Tuller, J. 1978. *The Scope of Hazard Management Expenditures in the U.S.* Working paper, Hazard Assessment Group, Clark University, Worcester, Massachusetts.

World Health Statistics Annual. 1976. World Health Organization, Geneva.

2

Risk and Comprehensive Planning

Man cannot completely train the great forces of nature; nor can he perfectly control his own processes or products. In the past, terrible disasters with great loss of life were accepted as inevitable. Today, greater awareness and expectations challenge what is acceptable, not only in health and safety, but also in pollution and amenity.

Rodin (1978)

Risk and the Comprehensive Plan

Communities are rarely prepared to leave their future to chance. With increasing frequency they attempt to ensure a desirable futurity by comprehensive planning which provides a guide for the area's development. In this way, only new infrastructure is built which permits a variety of social goals, such as mobility, employment, and education, to be met. Comprehensive plans, however, are not static and are periodically evaluated and revised to allow the community to deal with unforeseen changes in circumstance. For example, with the recent increase in energy costs, some cities have included efficiency of operation as a major goal in their planning. Future decisions will therefore be evaluated against this objective and energy conservation will be given a status that it has long been denied.

Normally, once a comprehensive plan has been approved by a community's elected officials, it is used as a basis for the development of land use zoning, housing plans and programs, the identification of new major streets and transportation services, selection of areas to be serviced by utilities, and open space acquisition. Since each of these activities influences community risk, it is apparent that if disaster frequency is to be reduced, then safety must also be sought as a major goal in comprehensive planning.

In most cases the final responsibility to ensure that the comprehensive plan is implemented rests with elected officials. However, these individuals rarely make significant decisions without expert advice, provided by civil servants, professional panels, and citizens' organizations. Many such advisory bodies have a single mandate, they are required to monitor the progress being made toward meeting one of the goals of the comprehensive plan. If the frequency of disaster is to be lowered, then most local governments require the assistance of such a committee, dedicated to monitoring risk and improving safety levels. This body should seek to ensure that the impact of existing hazards is reduced and that new developments do not increase community risk. Since the spectrum of threat is rapidly increasing, it is important that no single department is given this responsibility, but that the safety committee includes individuals with a wide variety of expertise. Such an advisory group to the council or regional board is necessary because local authorities cannot simply rely on federal or regional governments to monitor risk and to implement safety measures. The task is too great and hazard identification and reduction at international as well as national levels must be supplemented by local activities.

Since birth defects are often an early indication of increased general danger, an obstetrician should be invited to participate as a member of the safety committee. There are, for example, significantly more malformed children in Shawinigan, Quebec, where polyvinyl chloride plastic is manufactured than in Drummondville, a town of similar size in the same province (Theriault, 1977). Animals are also frequently more subject to environmental poisons than humans. The "cats dancing disease," a reckless and frenzied spinning, was the first sign of the mercury poisoning which, when it manifested itself in the fish-eaters of Minimata, Japan, in the 1950s, killed at least 90 people and left thousands crippled, blind, or paralyzed (Hutchison and Wallace, 1977). The committee would, therefore, benefit from the membership of a veterinarian. Similarly, the body tissues of bees take up and concentrate impurities. In general, the levels of contaminants on, or in, these tissues exceed the levels of these substances in air, flowers, pollen, nectar, or water. Toxic substances collected when foraging may poison large numbers of bees or shorten lifespans and lead to a slow but steadily diminishing population from chronic poisoning. Since they magnify contaminants, bees are a useful indicator of changes in environmental quality (Bromenshenk, 1978) and an active apiarist would also be a useful safety committee member. The populations of many birds, such as the falcon, may show early signs of environmental poisoning and an ornithologist or a "birder" might also be included.

Industrial workers or those involved in transportation are subjected to new hazards early in their production cycle. Refuse disposal often leads to exposure to dangerous wastes. It would appear logical to approach the unions involved to appoint suitable members to the safety committee. This body should also include geologists, meteorologists, biologists, engineers, chemists, physicists, and representatives of any state or federal disaster agencies operating in the area. Local health officers should also be appointed to serve.

If the safety committee is to ensure that community risk is reduced then it should have an active coordinator. This individual, to be effective, must have sufficient stature in the public service and the community to command the respect and obtain the cooperation of those with whom he will deal. Adequate engineering, secretarial, and other support services must also be made available. The safety committee coordinator should report directly to the mayor and council or to a very senior level of government to ensure proper backing for their activities. They should also establish links with other government agencies, professional groups and industries that will be of assistance during the preparation and implementation of the safety aspects of the comprehensive plan and its related documentation. There will inevitably be overlap with the legitimate concerns of departments involved with zoning, and with commissions or committees with a mandate to promote conservation and environmental impact assessment. For this reason, coordination of activities and interaction with such bodies is essential.

The Development of a Community Safety Plan

Hazard Identification and Assessment

While the comprehensive plan is normally the key document guiding community development, it rarely stands alone. Such a blueprint is usually a synthesis of a variety of more specific plans which have focused upon a single aspect of need. Other studies, for example, may have examined transportation objectives or educational alternatives in considerable depth, only their major conclusions being incorporated into the comprehensive plan itself. Risk too requires similar detailed analysis. A newly appointed safety committee should have the preparation of such a plan as one of its top priorities. An outline that might be followed in its compilation is presented in Figure 2.1. The major aim of this documentation would be to allow risk reduction to be incorporated more fully into future revisions of the comprehensive plan.

The first step in the production of a viable safety plan is the identification and assessment of the hazards that exist in the region. To this end a wide range of literature must be reviewed, data collected, and interviews conducted. In addition to identifying existing threats, the safety committee would also assess planned developments, perhaps providing advice on risk reduction to existing environmental impact review bodies. Its main concern would be how such structural change might influence community risk. The safety committee would also be expected to act in a monitoring capacity, reacting rapidly to any unanticipated changes in environmental quality. Its work should be widely publicized and a mailing address and telephone number listed prominently in the directory so that the public could provide it directly with relevant information.

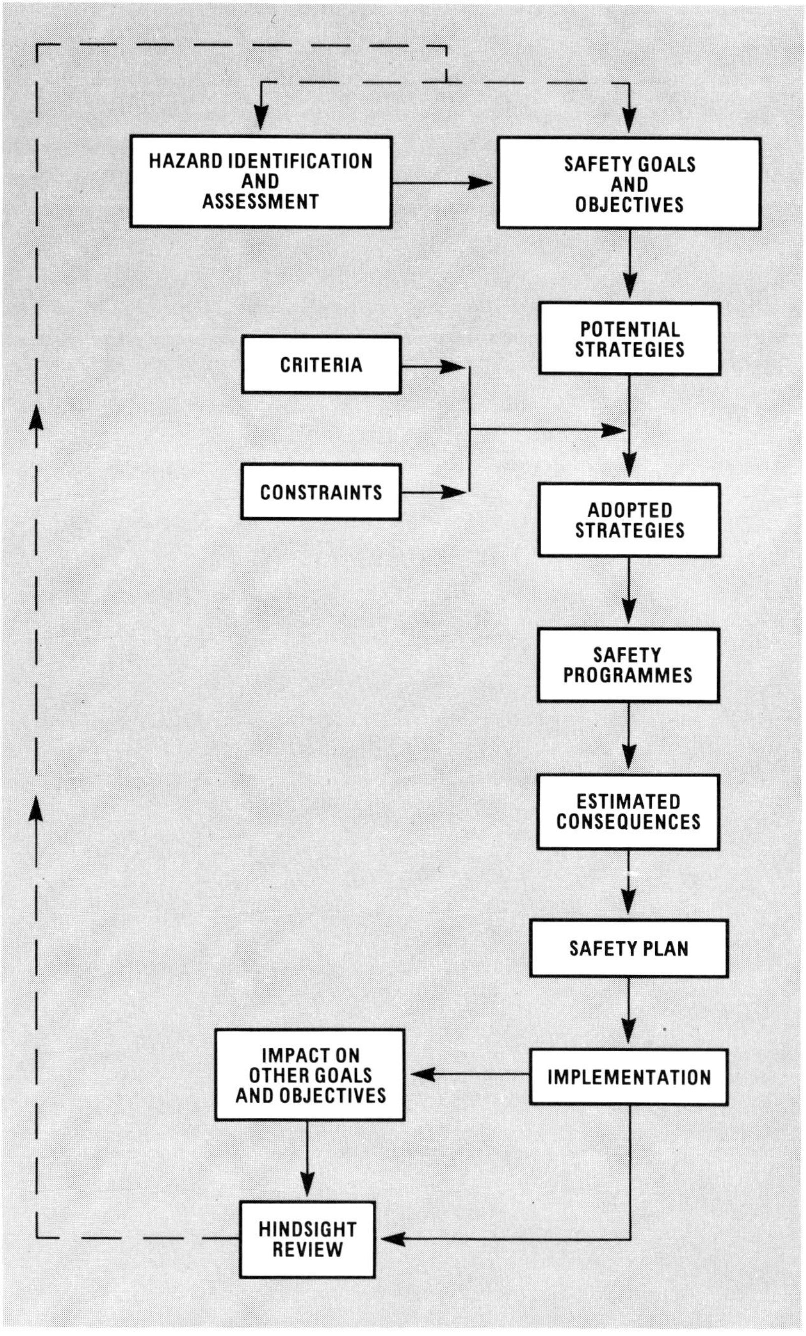

Figure 2.1. The development of a community safety plan.

PLATE 1. Failure of Tip No. 7, Aberfan, Wales, October 1966, resulted in the death of 144 people including 116 school children (National Coal Board photograph).

Safety Goals and Objectives

Quantifying Losses

Once the hazards that threaten a community have been identified, the losses for which they are responsible must be determined. Until this has been achieved, it is impossible to establish meaningful safety goals for in-

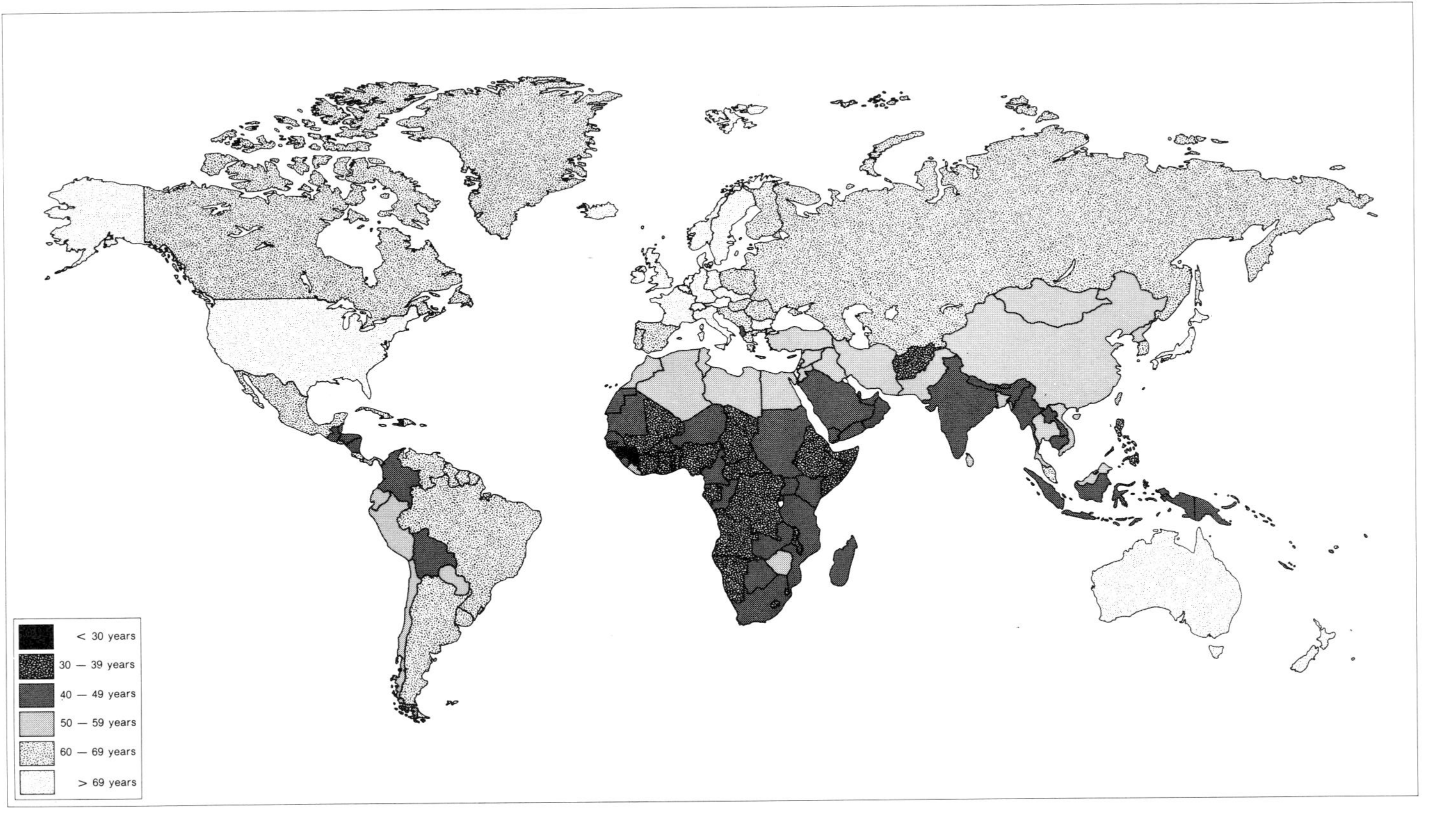

Figure 2.2. Expectation of life at birth.

corporation into the comprehensive plan since there is no baseline from which to measure achievement. Fortunately this task has been simplified, in many countries, by the mandatory recording of causes of death. Medical statistics covering many years often exist. These data allow regions or communities to establish how their mortality rates compare with those of other areas. Similarly, changes in the causes of death through time can also be detected in a single community.

Important sources of data on mortality include the statistics annuals of the World Health Organization and publications of the National Center for Health Statistics in the United States. In its simplest form this information can be displayed cartographically to illustrate world or national mortality patterns (Figure 2.2). Murray (1967), for example, mapped the rate of death from all causes in the United States on a countrywide basis. Howe (1979) has produced a comparable map showing mortality for males between the ages of 15 and 64, in the period 1970 to 1972, in the London boroughs.

At a more sophisticated level such mortality data are of value in allowing comparisons of death rates from individual diseases, hazards, or fatal accidents. Howe (1970), in his revised *National Atlas of Disease Mortality in the United Kingdom*, for example, used the standard error to detect significant local departures from the British norm. He could then display death rates and their relative statistical significance on the same map. When mapping mortality in the London boroughs he also employed a Standardized Mortality Ratio. This made allowances for variations of age structure and sex distribution within local populations as compared with the British norm (Howe, 1979). He then produced maps showing differences in mortality ratios for suicide; ischaemic heart disease; cancer of the trachea, lung, and bronchus as well as chronic bronchitis (Figure 2.3). These illustrated that there were significant differences between the London boroughs for these causes of death. Such maps generally established a decline in mortality ratios outward from the city center.

Not only do analyses of mortality data permit spatial differences in distinct causes of death to be established, but they can also allow changes through time to be identified. Monmonier (1974), for example, used standardized mortality ratios to compare the number of resident deaths from diabetes mellitus in New York State on a countywide basis, for the period 1957 to 1969. As seen from Figure 2.4, he was able to identify what he termed composite disease regions. In these it was possible to show if the rate of death from this illness was relatively high or low and whether it was increasing, stable, or decreasing. Not only should any comprehensive plan mortality goal establish a basis for setting targets for reducing adult death rates, it should also aim to reduce infant mortality. Statistics on this problem are also widely available.

One important deficiency in mortality statistics is that they often fail to fully account for risk of high mortality from rare events such as tornadoes,

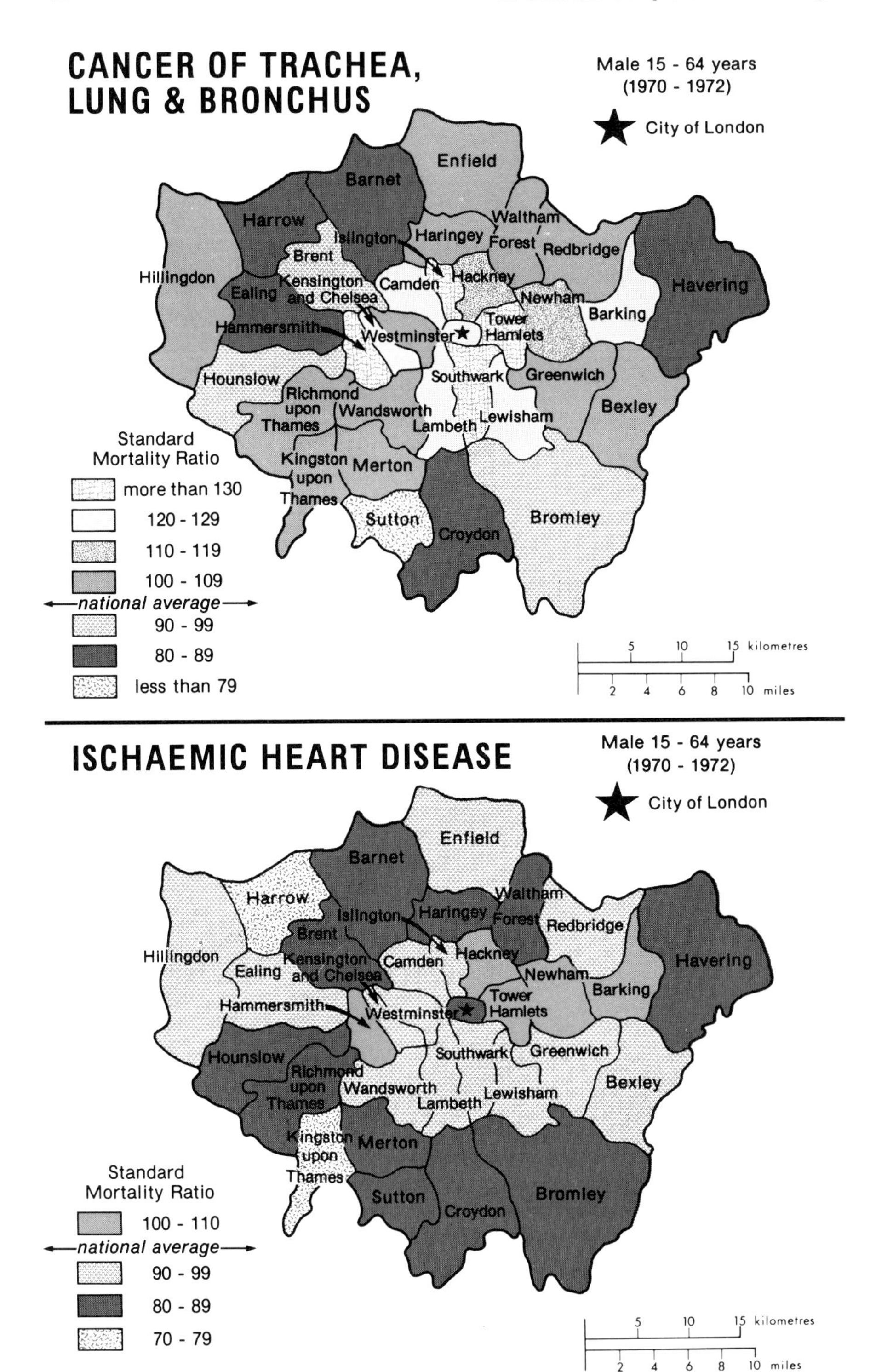

Figure 2.3. Causes of death in the London boroughs (Howe, 1979).

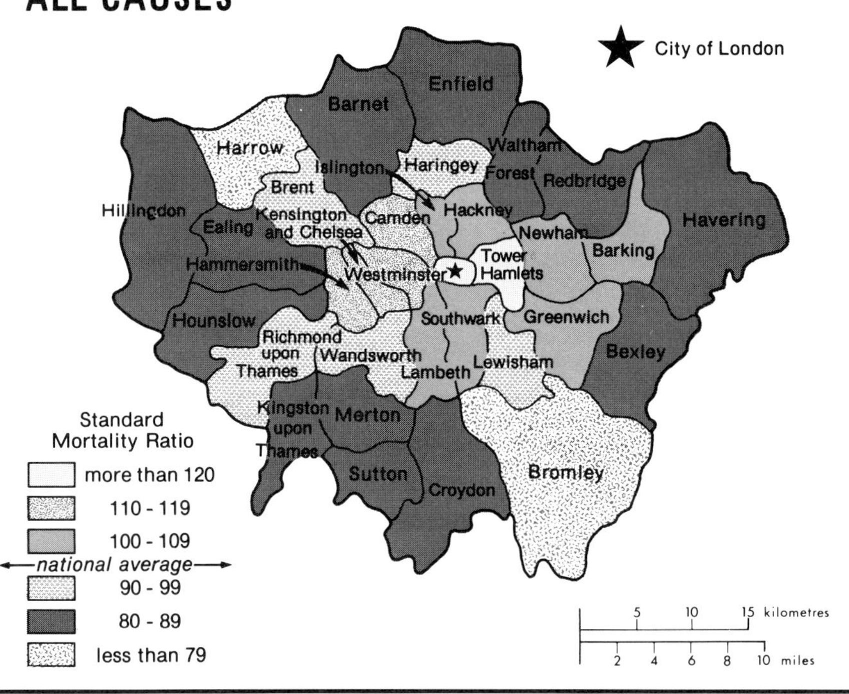

CHRONIC BRONCHITIS

Male 15 - 64 years
(1969 - 1973)

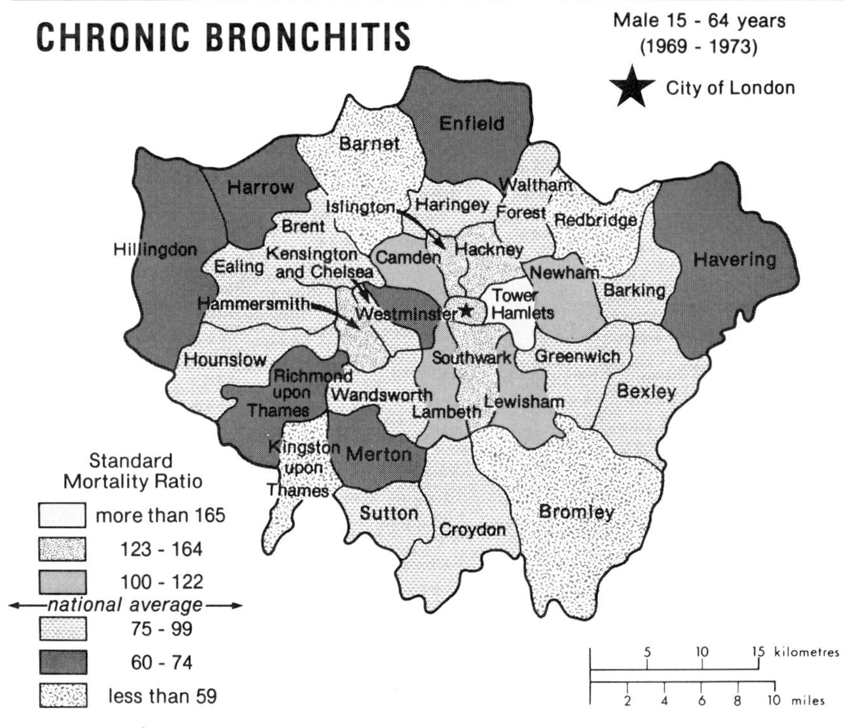

Figure 2.4. Changes from deaths from diabetes mellitus in New York State (Monmonier, 1974).

flash floods, or earthquakes. For this reason it is important that simulations of potential mortality from such disaster agents be carried out before strategies to meet mortality goals are finalized. Otherwise the potential impact of relatively rare, large scale destructive events may be overlooked.

Community safety plans and associated programs should also seek to reduce suffering from illnesses that are not necessarily fatal. This again involves identifying spatial differences in their incidence and a ranking of their local significance. Data should be collected on such problems as industrial accidents, birth defects, disability days, acute condition short-stay hospital discharges, and physician visits. This information is already available in some countries, including the United States where it is published by the U.S. Department of Health, Education and Welfare (1974) by geographic region, large metropolitan area, and other places of residence. Such information allows a general picture of the health of the community to be established by the safety committee and attempts can then be begun to identify those factors which appear to be reducing life quality and expectancy.

In addition to the deaths and illnesses associated with natural and man-made hazards, enormous financial costs are also incurred as a result of their impact. The safety goals proposed by the committee should also seek to reduce this toll. There is rarely any central agency monitoring hazard losses on a comprehensive basis. Nevertheless, numerous potential data sources exist. These include the records of insurance companies, fire and police agencies, and government departments of disaster mitigation, economic analysis, urban affairs, transport, and agriculture.

Establishing Safety Goals and Acceptable Risk

Once the morbidity, mortality, and economic loss rates from hazards have been established, a second step in goal setting must be taken, that of deciding upon acceptable safety levels. This process requires an answer to the question: How safe is safe enough? It will inevitably lead to difficult value judgments, conflicts, and trade-offs between other societal goals. Decision makers will have to face up to a variety of fundamental issues such as: How can the need for energy from nonrenewable resources be balanced against deaths from mining accidents or diseases caused by polluted air? To what extent do new health standards threaten employment or increase inflation? Will bans on hazardous insecticides reduce farm production? Such trade-offs are not new to planners, however, and are made between many other more traditional social objectives.

Similarly, problems arise because the public's tolerance of risk is uneven and varies widely among groups, technologies, and activities. As Fischhoff, Hohenemser, Kasperson, and Kates (1978) have pointed out, Americans accept 50,000 highway fatalities annually but demand extremely rigorous safety standards for nuclear power generation. There is pressure to spend as much as $3.5 million to save individual lives in the textile industry by lowering allowable concentrations of acrylonitrile in the air. In contrast, only a fraction of this is earmarked to prevent deaths from the wastes of fossil-fueled plants or liquefied natural gas facilities (Behr, 1978). Indeed, U.S. Health and Welfare Secretary Joseph Califano has called smoking a national tragedy that kills as many as 346,000 Americans a year, at a cost to taxpayers of $18 billion. Smoking is the largest preventable cause of death in the United States, yet it is a completely voluntary activity that, to nonsmokers at least, appears to have few, if any, associated benefits. This inconsistency of public values toward individual hazards complicates the design of a safety plan and the establishment of acceptable safety goals for use in comprehensive planning. Legal guidelines are often of little use, usually stipulating that risk should be "as low as feasible" or "as low as reasonably achievable" or "not unreasonable considering the benefits" (Fischhoff, Hohenemser, Kasperson, and Kates, 1978).

Nowhere on the earth's surface is free from hazards. Even if all possible steps have been taken to increase safety, some risk inevitably remains. If this is the case, then it appears logical for a community to set maximum allowable standards for geophysical, meteorological, biological, and industrial risks for differing activities. Through the comprehensive plan they should then guide the most significant developments to the areas where they will cause or sustain the least probable damage. The necessity for specific, quantitative standards has been recognized in a wide variety of planning fields, for example, in the management of air and water quality. Risk standards are, however, not strictly analogous. This is because risk is not a physical quantity and what constitutes an acceptable level is a psychological problem. There is really no such thing as a generally acceptable level of risk for involuntary activities. However, the converse is not true; society frequently judges some risks as being unacceptable. These social decisions are reflected in legislation dealing with health, welfare, and safety, including research funding for medicine, disaster mitigation, and building codes. The setting of risk standards, therefore, requires at a minimum making explicit the judgments of unacceptable risk which are demonstrated implicitly by society's activities (Puget Sound Council of Governments, 1975).

There is a growing literature dealing with the acceptability of risks. Safety committees would be well advised to read the 1977 publication of this name, produced by the Council for Science and Society and *Of Acceptable Risk* by Lowrance (1976). This literature demonstrates that there are at least four alternative approaches currently available for determining the level of acceptable risk. These will have to be reviewed before safety goals are set. The first of these, *risk aversion,* is unlikely to be adopted as a yardstick for measurement since it is most applicable to certain sophisticated technologies or highly publicized diseases such as botulism. It involves a decision to achieve the maximum possible reduction of risk, regardless of the costs involved. It does not allow comparison of the particular risk with others, or any balancing of benefits and expenses incurred. From the point of view of comprehensive planning such an approach is impractical since other goals and risks must be taken into consideration when budgeting.

A second method of determining what risks are acceptable and hence in setting goals for community mortality, morbidity and economic loss is that of *risk balancing*. This approach assumes that some level of risk above zero is socially acceptable. It tries to determine what this is by comparison with appropriate reference cases, such as similar technologies, natural background levels, or risks from other activities.

Normative rationality is a valuable concept in discussing what are acceptable levels of risk to life. While the concept of freedom must entail a degree of right to risk one's own life and limb, most people would still judge that mountain climbers, film stuntmen, and skydivers are taking unnecessary risks. Clearly, such intuitive standards can be operationalized and made explicit. All urban developments or technological innovations are associated with risk to life or health. Over relatively long time periods, society moves toward establishing what is a permissible level of risk for any such system. This is done by allocating private and public resources to increase the level of safety to the point at which it is publicly acceptable. One method of determining the acceptable level of risk from all hazards is to establish the standards which have been applied to current urban-technological systems (Puget Sound Council of Governments, 1975). This is known as the revealed preference method of approaching the issue.

Starr (1969) has undertaken such a quantitative analysis of the accidental deaths arising from the use of modern technologies by the public. This was done in an effort to provide some understanding of the public's approach to balancing the economic and social benefits of innovation against its associated risks. Starr derived what appeared to be four laws of acceptable risk. The first of these was that the acceptability of risk is roughly proportional to the cube (third power) of its associated benefits. Second, it was concluded that the public appears willing to accept risks from voluntary activities such as skiing or hang-gliding which are approximately 1000 times higher than it would tolerate from involuntary activities that generate comparable benefits. An individual may, for example, move to the suburbs to avoid higher property taxes and accept the associated, increased risk of being killed in a freeway traffic accident. This is a voluntary risk, which contrasts with involuntary risk, for example, that due to the activities of government or private corporation over which the individual has no control. Such risks would include those of being killed by the failure of a hydroelectric dam or leakage from a chemical factory.

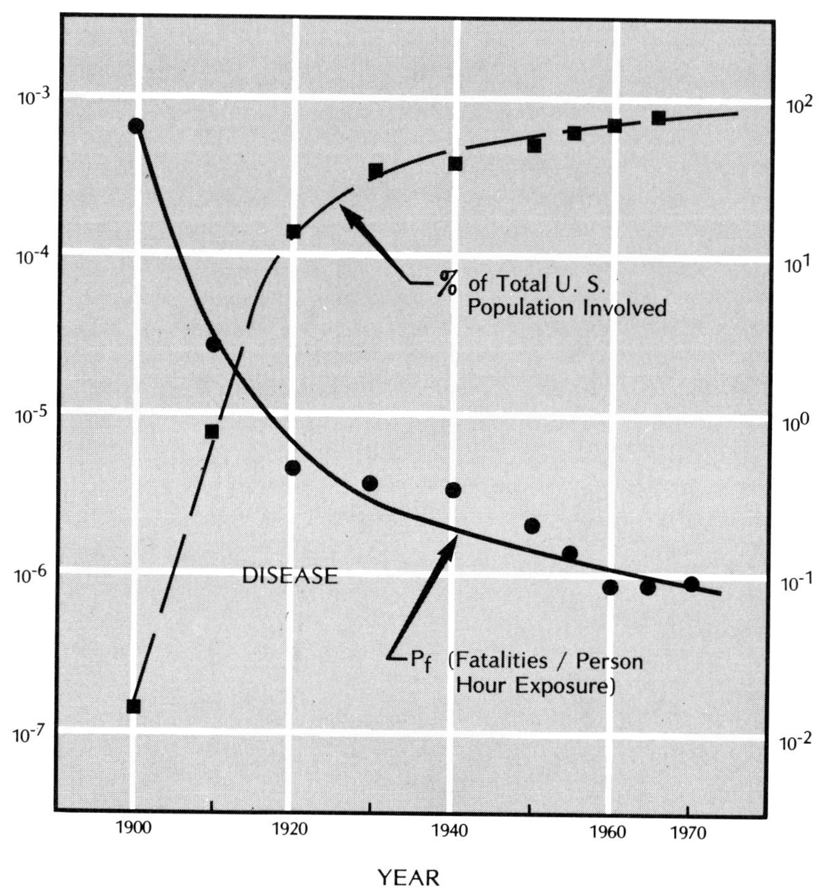

Figure 2.5. Risk and participation trends for motor vehicles (after Starr, C., 1969. *Science*, **165**: 1232–1238. Copyright 1969 by the American Association for the Advancement of Science).

Third, Starr's research appeared to show that the acceptable level of risk is inversely proportional to the number of individuals exposed to it (Figure 2.5). Finally, he concluded that the level of risk tolerated for voluntarily accepted hazards, such as sports, is similar to that from disease. On the basis of this last observation Starr postulated that the rate of death from disease appears to play, psychologically, a yardstick role in determining the acceptability of risk on a voluntary basis. Figure 2.6 depicts the results of Starr's analysis, while Table 2.1 illustrates average risk levels for differing activities. Starr's work has been subject to criticism. More extensive research by Otway (1975) did not confirm these regularities. Moreover, it has been pointed out that at high level of risk where death or serious injury is relatively probable, an individual must seriously ponder his fate and a simple algebra of costs and benefits is inadequate (Council for Science and Society, 1977).

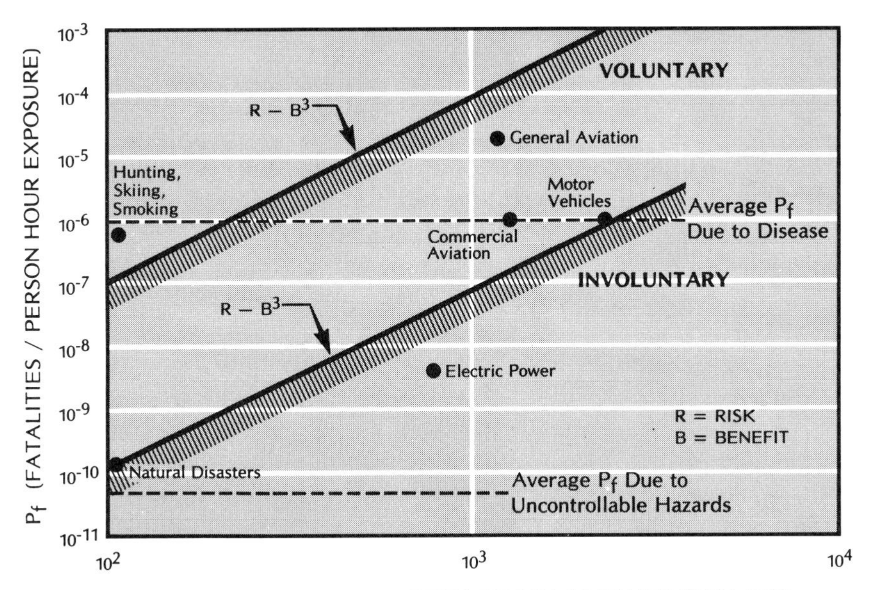

Figure 2.6. Risk compared against benefits (after Starr, C., 1969. *Science*, **165**: 1232–1238. Copyright 1969 by the American Association for the Advancement of Science).

Nevertheless, Starr's research has some value. Evidence indicates that the long-term societal adjustment to the involuntary risk of death from technological systems seems to approach the average natural disease death rate at the upper limit of benefits. It can perhaps be argued, therefore, that any higher risk levels are unacceptable in Western society. In contrast, the average risk due to natural disaster might be taken to represent the minimum "background" level of risk. The risk of death from

Table 2.1. Average Life Risk Levels[a]

Activity or event	Fatalities per person-hour of exposure
Natural disasters	1 in 100 billion (10^{-11})
Fossil power plants	2 in 10 billion (10^{-10})
Radiation (100 mrem/yr)	5 in 10 billion (10^{-10})
Electricity	1 in 1 billion (10^{-9})
Smoking	5 in 10 million (10^{-7})
Hunting	9 in 10 million (10^{-7})
Skiing	9 in 10 million (10^{-7})
Commercial aviation	1 in 1 million (10^{-6})
Motor vehicles	1 in 1 million (10^{-6})
Natural disease	1 in 1 million (10^{-6})
General aviation	3 in 10 thousand (10^{-4})

[a]Data from Starr (1969).

earthquakes, for example, has been calculated at approximately one fatality per 100 billion person-hours of exposure. Naturally this varies with the location and design of building. A safety committee might consider suggesting the use of these two limits to acceptable risk, of death from disease and from natural hazards, to set standards for specific activities and control land use. By modifying this technique, safety goals might be set with reference to national mortality, morbidity averages, or safety levels achieved in more risk-conscious countries or communities. A municipality might set the safety goals of its comprehensive plan so that losses of life and injury from a variety of diseases and accident types and hazard-induced economic costs were all below the national or regional average.

A third possible method that might be used to set community safety goals involves determining acceptable risk by using *cost effectiveness* techniques. Such an approach seeks to maximize the reduction of risk for each dollar expended on safety measures. Acceptable risk can be established by breaks of slope of risk reduction efficiency for a given hazard. That is, additional expenditure will cease when it results in significantly smaller increases in loss reduction. Acceptable risk can also be determined by allocating the available funds among hazards so that the expenditure results in the greatest reduction in risk for society as a whole. If this economic technique is applied, then once the rate of hazard-related losses has been established, each disease, illness, or cost should be studied to determine where the greatest reductions can be achieved for the minimum investment. It should be remembered, however, that some reductions of risk can be accomplished at little or no cost, through a change in practice or process, not necessarily more expensive than that already in use. Prohibiting smoking in office buildings provides an interesting example.

A final method of determining acceptable risk is termed *cost-benefit balancing*. This method recognizes that some level of risk above zero must be accepted if other objectives are to be met. Just what this level is, and how rigorous are the safety goals which will be set, is decided by balancing the benefits of an activity against the level of risk it entails. To illustrate, the higher the degree of employment and profit associated with a technology, the greater the socially acceptable risk. Losses tolerated, therefore, increase proportionally with the size of the resulting benefits. This approach was taken by the authors of the disaster mitigation plan for the Central Puget Sound Region (Puget Sound Council of Governments, 1975). They argued that the higher the locational benefits accruing to their region from an activity, the greater the acceptable level of associated risk. Locational benefits were defined as "the total social benefits accruing to the region less the benefits which could be gained by locating the activity outside known hazard areas." The implied differences in benefits are caused by extra distance to markets and transportation routes, and sunk costs such as those of site preparation. When benefits are measured in this manner, those activities with the most restrictive siting requirements, such as port facilities and river gravel extraction, also have the greatest locational benefits. This follows because these industries would be entirely precluded if they were not allowed to locate on a very narrow range of sites. It was therefore argued that the more restrictive the locational demands of a particular land use, the greater the risk to life that should be tolerated.

Potential Strategies

Once current health and economic loss levels due to community hazards have been established and goals set for their reduction, determined attempts must be made to understand why such phenomena occur. This knowledge is an essential first step in designing strategies for land use, operating procedures, and design criteria which will be required if safety targets are to be met.

This task may not be simple. The guiding philosophy in medical science has long been the notion of specific etiology, the "one germ:one disease" concept propounded by such physicians as Louis Pasteur and Paul Ehrlich in the late nineteenth century (Howe, 1979). This thesis has been proved repeatedly by the discovery that pathogenic microbes, viruses, bacteria, and the microscopic life-forms were responsible for such common ailments as malaria, diphtheria, and tuberculosis. This concept, that every human illness had a particular cause, has led to the invention of a variety of wonder drugs capable of controlling or curing numerous afflictions. Many of those who die in the Developing World still do so from such illnesses and establishing the distribution of such mortalities may quickly lead to a clearer understanding of their ultimate cause. This approach was successfully used as early as 1854 during a cholera epidemic in London. Snow plotted the distribution of deaths and found them to be concentrated in the vicinity of the Broad Street Pump, clearly suggesting that the disease was being dispersed by polluted water (Stamp, 1965).

In the case of afflictions of the Developed World such as cancers, mental illness, and cardiovascular disease, there are no single causes. These problems are known to have "risk factors" or "risk indicators" associated with them and appear to be the result of ecological imbalance or maladjustment in the overall milieu (Howe, 1979). They are probably a response to environmental hazards or stimuli, conditioned by the individual's genetic inheritance. Exactly what combinations of life style factors or environmental variables predispose to, or account for, chronic bronchitis or strokes is still uncertain. However, by mapping morbidity and mortality statistics it is apparent that each of these diseases has its own specific distribution pattern. Sometimes the spatial patterns of a disease suggest

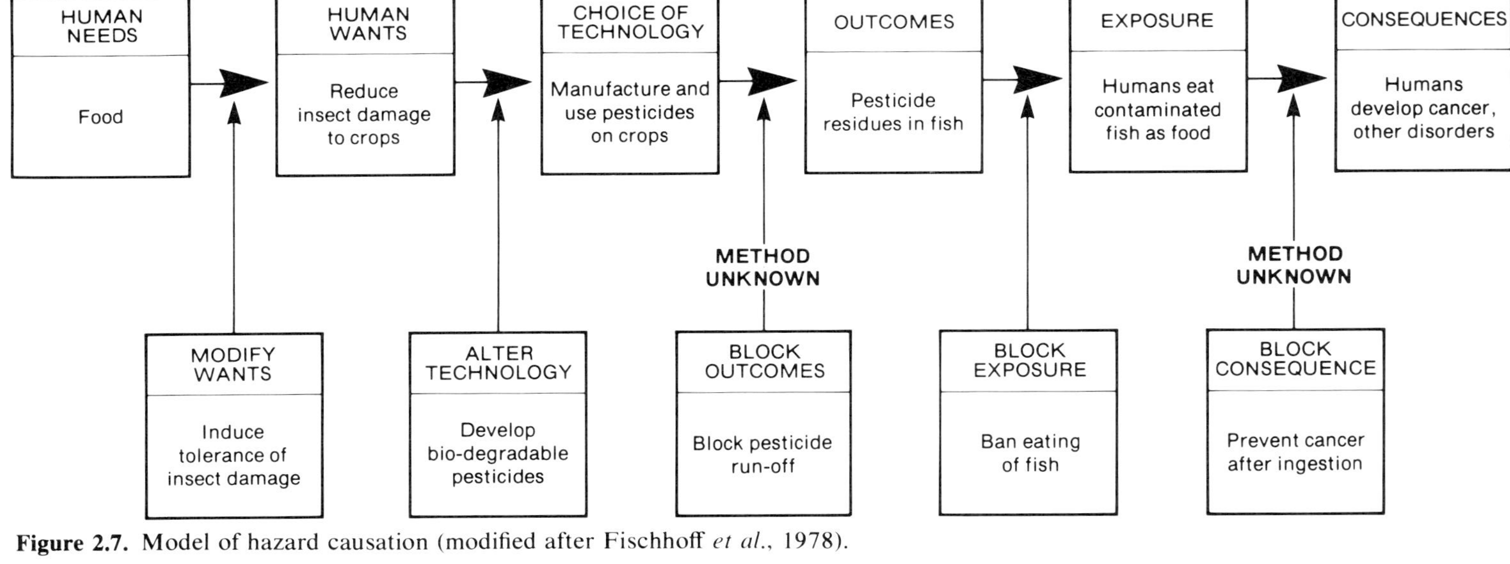

Figure 2.7. Model of hazard causation (modified after Fischhoff *et al.*, 1978).

illustrates the various levels at which preventative strategies might be applied. Higher-order events could be removed by policy changes, for example, driving and/or drinking ages could be raised. Initiating events could be attacked, as exemplified by compulsory annual vehicle inspection, roadside police checks for drunken drivers, or the segregation of traffic and pedestrians by better highway construction. Outcomes might be modified by improved automobile design to reduce the destructive nature of impact, while an upgrading of ambulance and medical facilities might mitigate adverse consequences.

The first logical step in the design of safety strategies is the listing of all major hazards identified in the region. Many of these will be common and the literature will contain careful assessments of how best to reduce their adverse impacts. Others will be unique, or nearly so, to the region and it is these that should then be subjected to the modeling process described above, so that the sequence of events, outcomes, and consequences that lead to hazard-related mortalities, morbidities, and economic losses can be recognized. Once such information has been collected for each hazard, a wide range of strategies, capable of reducing such adverse events, can then be recognized and evaluated. It is also the responsibility of the safety committee to take an active part in the evaluation of new development schemes. To a large degree, environmental impact statements or their equivalents are essentially attempts to compare the benefits of development with the risks these create. Safety committees have much to add to this debate.

Criteria for Evaluating Alternative Strategies

The identification of the most desirable alternative strategy for reducing risk is largely a process of learning from the experience of others. Few hazards are unique, most safety committees can gain valuable insights from hazard-related literature, interviews with disaster survivors, and contacts with those involved elsewhere in mitigating community risk. In this way a variety of strategies for reducing losses from particular hazards can be identified, such as those shown in Table 2.2 for floods. These must then be evaluated against a series of criteria to allow the selection of the most desirable. The final choice of loss reducing strategies is political and will eventually depend on the weight placed on safety by elected officials as compared with the emphasis given to other goals that the society is also attempting to achieve, such as economic growth and environmental quality. Table 2.3 includes 14 criteria that might be used in this evaluation process.

The assessment of individual strategies against these 14 evaluative criteria allows a ranking of the basis of desirability. There are, for example, a multiplicity of alternative ways of reducing flood losses (Table 2.2). However, the application of this method of comparison might lead to adoption

Table 2.2. Potential Adjustments to Floods, Appropriate Applications, Special Features, and Implications for Policy Making

Adjustment	Appropriate applications	Special features	Implications for policy making
Flood protection	Especially appropriate where the potential losses are large, and particularly where the flood plain is highly urbanized and industrialized.	Flood protection works can often be provided as part of a multiple purpose project, such as a storage reservoir.	Tends to encourage persistent human occupance and may create a false sense of security among flood plain dwellers. Tends to shift responsibility from flood plain dwellers to the public at large. Needs to be supplemented by other measures.
Watershed management	Enough runoff remains in the low water period after the programs have been undertaken.	Can accomplish multiple objectives by involving forestry and/or agriculture. Is a useful complement to other measures	Most of the benefits are derived close to the area where watershed mangement is undertaken.
Weather management	Where external or side effects are minimum and those adversely affected can be compensated	Still in a largely experimental stage with overall effects still somewhat uncertain	Needs governmental control to ensure that the public at large is protected and that losers are compensated by gainers
Flood proofing	To scattered buildings that are frequently flooded, and especially where the depth of flooding is less than 3 feet, and more than 3 hours warning is possible.	Structural measures which can be combined with other measures such as emergency action or land use regulation.	Requires a well-organized flood forecasting system, aided by a flood hazard informaton program Depends largely on individual action and tends to lag in intervals between floods.

Table 2.2. (*continued*)

Adjustment	Appropriate applications	Special features	Implications for policy making
Flood plain mangement	Where uses other than agricultural and recreational uses are competing for flood plain land, and especially where such uses involve urban and industrial uses.	Combinations of structural methods and nonstructural methods.	Tends to encourage efficient use of flood plain land. Responsibility is shared between the flood plain dweller and the regulatory authorities. Requires a rational basis for selecting land uses, and strong leadership and public support to ensure adoption.
Emergency measures	Everywhere, but especially where the flood-to-peak interval is greater than one day, the flood duration is short, flood frequency is high.	A nonstructural measure which can be used independently or in combination with other measures	Requires a well-organized flood forecasting system with clearly defined and announced responsibilities for these functions. Tends to encourage persistent human occupance. Interest tends to lag when flood frequency is low.

Source: Sewell and Foster (1976).

of land use regulations and zoning ordinances on the basis of their relatively low cost to government, ease of administration, and dearth of environmental impact. They can also normally be applied without reference to higher levels of government.

Constraints and Strategy Adoption

The management of hazards does not take place in a vacuum. In consequence, many political and socioeconomic factors limit the number of actual strategies which can be applied by any society to reduce losses. Seven such significant constraints have been identified by Fischhoff *et al.* (1978). They are the lack of sufficient knowledge to link causes with ef-

Table 2.3. Criteria for Evaluating Mitigation Strategies

Criteria	Strategy-related questions
1. Equity	Do those responsible for creating the hazard pay for its reduction? Where there is no man-made cause, is the cost of response fairly distributed?
2. Timing	Will the beneficial effects of this strategy be quickly realized?
3. Leverage	Will the application of this strategy lead to further risk reducing actions by others?
4. Cost to government	Is this strategy the most cost-effective or could the same results be achieved more cheaply by other means?
5. Administrative efficiency	Can it be easily administered or will its application be neglected because of difficulty of administration or lack of expertise?
6. Continuity of effects	Will the effects of the application of this strategy be continuous or merely short term?
7. Compatibility	How compatible is this strategy with others that may be adopted?
8. Jurisdictional authority	Does this level of government have the legislated authority to apply this strategy? If not, can higher levels be encouraged to do so?
9. Effects on the economy	What will be the economic impacts of this strategy?
10. Effects on the environment	What will be the environmental impacts of this strategy?
11. Hazard creation	Will this strategy itself introduce *new* risks?
12. Hazard reduction potential	What proportion of the losses due to this hazard will this strategy prevent? Will it allow the safety goal to be reached?
13. Public and pressure group reaction	Are there likely to be adverse reactions to implementation?
14. Individual freedom	Does the strategy deny basic rights?

fects, an unwillingness to forego associated benefits, a limited capacity to react, divergence between perceived and actual risks, conflicting goals, institutional weaknesses, and the potential for creating new hazards.

There are many reasons why incomplete knowledge must inevitably restrict hazard management strategy selection. In a world of constant change, innovation is commonplace. Since many technologies are new and untried, safety systems must be installed without background knowledge about their effectiveness. As a result, there is often far too little information about the consequences of failure or of potential synergistic effects. Under these circumstances it becomes impossible to determine realistically whether or not such innovations exceed acceptable levels of risk to the community. Indeed, even though the design of a nuclear reactor, natural liquid gas transport facility, or dam may remain the same, any increase in warfare, terrorism, or the widespread adoption of some other innovation at a later date may substantially alter the risks involved (Council for Science and Society, 1977).

In addition, many harmful effects are specified through dose–response relationships. It is normally assumed that risk increases with the degree of exposure. As a rule, most standards are set by assessing the effects of high frequency, large doses on laboratory animals or by studying the results of the constant exposure of workers in a related industry. The consequences of low doses and occasional exposure is often unclear but is generally estimated by linear extrapolation from these data. Unfortunately it is this type of exposure to which the general public is subjected. For many hazards it is not clear whether there is any safe threshold. The dose–response curve relating cigarette smoking to lung cancer demonstrates that none exists for this activity (Fischhoff *et al.*, 1978). This lack of knowledge, which prohibits absolute safety standards, must obviously hamper hazard management strategy selection.

The creation of man-made hazards and the acceptance of natural ones is almost invariably associated with benefits as well as costs. Typically, neither the risks nor the benefits are evenly distributed throughout society. For this reason, for some groups benefits may greatly outweigh risks although for others the reverse may be the case. Naturally this often leads to conflict and political compromise. It also accounts for the emphasis of hazard management strategies aimed at mitigating low order consequences rather than removing initiating events. There are many examples of disasters that have been caused or magnified by the unwillingness of certain groups to forego the benefits of their actions, despite enormous adverse consequences for others. This appears to have been the case in such diverse business transactions as the sale of Swiss avalanche tracks to summer tourists and of thalidomide to pregnant women (Fraser, 1966; The Sunday Times Insight Team, 1978). Where the links between cause and effect are not fully documented or widely accepted, resistance can be expected from pressure groups representing those who anticipate loss from risk-reducing strategies. In some cases this may

PLATE 2. Oil from the Arrow fouling the shoreline at Arichat Harbour, Nova Scotia. Society as a whole benefits from oil transportation; risks are greatest in coastal communities (Ministry of Transport photograph).

PLATE 3. Oilevator in operation cleaning the shore at River Inhabitants, Nova Scotia, following the wrecking of the Arrow (Ministry of Transport photograph).

prevent the successful application of the strategy, as occurred during the risk mapping program undertaken in Anchorage, Alaska, after the 1964 earthquake (Schoop, 1969).

The very complexity of risk is a constraint influencing its management. The multiplicity of hazards has been graphically summarized by Rabinowitch (1972):

> One day we hear about the danger of mercury, and run to throw out cans of tuna fish from our shelves, the next day the food to shun may be butter, which our grandparents considered the acme of wholesomeness; then we have to scrub the lead paint from our walls. Today, the danger lurks in the phosphates in our favorite detergent; tomorrow the finger points to insecticides, which were hailed a few years ago as saviors of millions from hunger and disease. The threats of death, insanity and—somehow even more fearsome—cancer lurk in all we eat or touch.

This characterization is not an exaggeration. In his discussion of predicting the impact of natural hazards, Blume (1978) identified 88 classes of such disaster agents. In addition, the American Chemical Society has registered some 4 million chemical compounds, 32,000 of which are already in commercial use (U.S. Environmental Protection Agency, 1977). It is unknown how many of these are potentially dangerous although there are currently some 2400 substances that may be causing cancer in the workplace (Christensen, 1976).

There are just too many hazards for society or its individual members to seek to mitigate. There is not enough money, time, or public interest to attend to more than a few major threats at any one time. Fischhoff *et al.* (1978) have shown, for example, that if the U.S. federal government were to conduct standard 500-rat, $250,000 studies of carcinogenicity on each of the 4 million compounds currently known, it would require 2 billion rats and a research budget of $1 trillion. In addition industry would typically have to spend some 20 times as much to enforce resulting government mandated safety and environmental regulations (Weidenbaum, 1978). Nevertheless, statistically speaking, a 500-rat study is inadequate. To quote Lowrance (1976), commenting on a United States Federal Drug Administration document on carcinogens:

> Even with as many as 1,000 test animals, and using only 90 percent confidence limits, the upper limit revealed by negative experiment (one revealing no tumors) is 2.3 cancers per 1,000 test animals. No one would wish to introduce an agent into a human population for which no more could be said than that it would probably produce no more than two tumours per 1,000. To reduce the upper limit of risk to two tumours per one million (at confidence limits of 99.9 percent) would require a negative result on somewhat more than three million test animals.

In summary, it is virtually impossible to test for carcinogenicity in a statistically valid manner for even one, let alone the 4 million compounds now available. Problems such as these which may require billions of test animals and millions of research workers to address have been termed trans-scientific by Weinberg (1972). That is, their solution is beyond the practical application of the scientific method, given the limits of human resources and finances.

There are at least two possible major responses to the myriad of hazards associated with high technology. The first is for hazard managers to pick and choose among a long and growing list of natural and man-made hazards, emphasizing strategies to reduce those that are known to cause the greatest losses. Other dangers will be neglected until they begin to be more dramatically reflected in mortality or morbidity tables. An alternative reaction to the threats associated with technology is to avoid its low order consequences by rejecting its initiating events. This approach to risk reduction is at least partially responsible for such diverse phenomena as the growth of the environmental movement, the lobby against nuclear power, support for renewable energy resources, the development of agricultural communes, and the spread of health food shops. Indeed, part of the flight to the suburbs is probably an attempt to escape the threats of the inner cities, such as violent crime and air pollution.

The view varies with the viewpoint. Yet it is the perception of hazards that shapes the legislation passed by politicians, the reactions of manufacturers, and the safety programs of professionals. That there is a divergence between actual and perceived risk has been demonstrated by numerous studies. Lawless (1974) reviewed 45 major public alarms over technology and found that in over 25% of the cases opponents of development had overestimated the threat, while in more than 50% of the examples examined proponents had underestimated it. Professionals tend to selectively identify those hazards that they have been specifically trained to deal with (Sewell, 1971; Barker, 1977). While the public tends to overestimate mortality rates from well-publicized hazards such as botulism, floods and tornadoes, it underestimates those from most chronic causes of death, such as diabetes, stomach cancer, and strokes (Fischhoff, *et al.,* 1978).

Such misconceptions govern how society and individuals respond to hazards and the types of adjustments that are made to reduce risk. One significant aspect of any safety program must be establishing just what hazards occur, and how they compare in their potential for death and destruction. Unless this information is generally accepted, misinformation will act as a constraint on the choice of hazard management strategies. As a result, safety programs will be designed to mitigate losses from perceived rather than actual threats.

Society pursues more than one goal. The need for safety must compete with others such as the desire for employment, food, water, transportation, and energy. In some instances strategies to reduce losses from a particular hazard may conflict with those designed to satisfy other national or local goals. Trade-offs must be made which place constraints on disaster mitigation programs. A recent example of this process was President Carter's intervention to weaken proposed regulations to protect some 150,000 to 800,000 workers from cotton dust disease on the grounds that the legislation was inflationary (Shabecoff, 1978).

Institutional weaknesses form a further major constraint on the adoption of optimum strategies to reduce risk. In some cases the same agency is responsible for both promoting and controlling particular types of development, as is the case with the Joint Committee on Atomic Energy in the United States. Alternatively, jurisdiction over a particular hazard may be divided between a variety of federal, state, or regional agencies. This is typically the case with many chemicals which are subject to different authorities and regulations when in food, air, water, or on the land. Under these circumstances, it becomes increasingly difficult to pursue safety measures that are widely applicable.

The full range of hazard control strategies is often not adopted because they themselves may produce additional risks. In Japan, for example, the Tokyo district court has just awarded damages of $119.7 million to compensate 133 victims of subacute myelo-optico neuropathy, a nervous disorder. Similar cases are being heard in 23 other Japanese district courts. This crippling disease has apparently been caused by the drug clioquinol, a substance used since 1899 in anti-diarrhea medicines. More than 11,000 Japanese have suffered from the disease in the past two decades (Chapman, 1979). Similarly, faulty X-ray equipment designed to aid in the discovery of ailments may be responsible for some cases of cancer. *Time* (April 2, 1979) reports that, in an unpublished report being circulated among United States government and industry aviation experts, the National Aeronautics and Space Administration analyzes 2856 confidential reports from pilots describing near midair collisions. Of these, 65% occurred in the vicinity of airports. There appear to be 11.9 such near misses for every million flights. Perhaps the most interesting finding of the survey would seem to be that the highest near crash rates take place in areas where pilots fly under the direction of ground controllers who are using sophisticated radar equipment. Apparently, concentration on the controller's signals diverts visual attention from the skies. It would seem that the more elaborate the safety procedure and equipment, the greater the probability of collision, the reverse of what was intended or anticipated. A further example of the creation of risk by the adoption of a strategy to remove it can be seen in the use of biodegradable pesticides to lessen threats to the environment which has exacerbated occupational hazards. Where the adoption of a hazard reduction strategy involves innovation it always carries with it the potential for unforeseen risk. This knowledge can act as a constraint in its selection.

Safety Programs

Once a safety committee has determined what hazards are adversely affecting the community, it is in a position to advise elected officials on how these threats can be reduced. In most communities hazard mitigating strategies are applied on an ad hoc basis in an uncoordinated manner. They are likely to be far more effective, however, if they are implemented in groups to form six interrelated disaster reducing programs. Such programs should be at the core of the community safety plan (Figure 2.1). Their implementation will greatly assist in the achievement of any safety goals set out in the comprehensive plan. Each of these safety programs is described in detail elsewhere in the volume. They involve determining the spatial distribution of risk, increasing safety through better design, disaster prediction, establishing warning systems, and the preparation of disaster and reconstruction plans.

The success or failure of every safety program rests on a valid appreciation of the distribution of risk. For this reason, specific and total risk maps must be prepared to determine how hazards differ in their spatial distribution and potential severity of impact. These must allow risk to be portrayed in a manner that will ensure its incorporation into the decision-making process. In this way activities can be sited so as to ensure minimum natural and man-made hazard losses. Risk is not an absolute but can be reduced further by improvements in building codes, highway design, safety by-laws and air, land, and water quality standards.

Warning systems are also key elements of all safety programs. These should be established to allow the monitoring of potential disaster agents and to permit evacuation or other preventative measures to be taken when danger reaches critical levels. Response to such warnings is inevitably most successful if community awareness has been heightened and a disaster plan drawn up to ensure that all necessary tasks are accomplished with a minimum of delay or confusion. Where destruction occurs, recovery is heightened by the availability of plans designed to speed rational reconstruction and minimize delay. The consequences of implementing such programs should be assessed against the safety goals that have been set by the community. Naturally they should be designed so that they permit these to be met.

Safety Plan

The six basic programs described above should, in fact, form the basis of the community safety plan, which can then be used to ensure that safety goals in the comprehensive plan are met. One fundamentally significant point, however, requires emphasis here. The six key program elements must be implemented in a logical, sequential manner as shown in Figure

2.8. This is because each program forms an essential stepping-stone for those that follow. Total risk mapping, for example, must precede simulation which is an essential prerequisite for the preparation of any successful disaster plan.

Much of the information required to prepare such a community safety plan, such as borehole records, geological surveys, and land use maps, are likely to be already available. What is required is the ability to synthesize this material into the required format. Once such a plan has been developed by the safety committee it must receive official sanction. As has already been described, perhaps the best way of achieving this is to include safety as a major goal in the area's comprehensive plan. The six safety programs can then be implemented as the measures which are required to achieve this objective.

Hindsight Review

The changing nature of risk and the multiplicity of potential strategies for reducing it mandate that hindsight reviews be conducted as an integral part of every safety plan. Progress toward the achievement of safety goals must be evaluated so that modifications of direction can be made wherever necessary. This process requires a constant monitoring of mortality, morbidity, and hazard-related economic loss. While these data in themselves may be sufficient to assure safety planners that risk is being reduced, it is useful to demonstrate this to the general public in a more easily understandable manner.

Monetary Safety Indices

One method of making citizens more aware of the value of risk reduction is to publish a safety equivalent of the consumer price index. Such an approach would involve assigning a monetary value to life loss and injury and adding the resulting total to hazard-related property damage. The final figure could then be modified to accommodate changes in population and released to the local press on a monthly basis. In this manner changes in per capita losses could be monitored.

Such an approach requires placing a price on human life. Linnerooth (1976) has identified six methods that have been used in assigning a monetary value to changes in population mortality. These techniques can conveniently be subdivided into two groups on the basis of how mortality is specified. The first group is composed of methods in which mortality risk is expressed in terms of expected lives lost and a fixed value is attached to each life. It includes the human-capital approach which values each life according to the discounted future earnings of the individual. This tech-

ACTIVITY	FIRST YEAR	SECOND YEAR	THIRD YEAR	FOURTH YEAR
RISK MAPPING	Ranking hazards; Identifying hazards; Data collection on occurrence	Establishing frequency and intensity; Single hazard maps	Establishing risk standards; Total risk map	Enforcing risk standards
GREATER SAFETY THROUGH BETTER DESIGN	Building data bank; Structural integrity survey	Security review; Identification of high risk buildings	Passage of abatement ordinances; Revision of building codes	Promotion of fail-safe design and forgiving environments
DISASTER SIMULATION AND PREDICTION		Computer simulations of disaster; Scale models	Delphi; Scenarios; Disaster games	Field exercises
WARNING SYSTEMS		System design	Education of users; Needs of special groups	Testing system
DISASTER PLANNING		Plan preparation	Communications; Evacuation preparations; Pre-impact preparations; Shelters	Revision; Control centre
PLANNING FOR RECONSTRUCTION			Spatial estimates of damage; Emergency zoning	Plans for rebuilding and employment; Potential revisions to building code

Figure 2.8. Major elements in a safety program.

nique has been widely used to calculate optimum life insurance (Dublin and Lotka, 1940), economic losses from accidents and disease (Reynolds, 1956; Fisher 1909), and more recently to provide a monetary value for the benefit of potentially saving a life through improving safety standards (Fromm, 1971; Ridker, 1970). Life values resulting from the application of this method have generally varied between $100,000 to $400,000. The United States National Center for Health Statistics has calculated, for example, that a millworker who dies in his late twenties loses $274,496 in lifetime potential earnings. This approach, however, has been highly criticized, and was termed "quantification gone mad" by Devons (1961), who wrote:

> Estimating the net loss to the community in this way leads to the fantastic result that if we could have more road accidents in which we succeeded in knocking down and killing old people we should reduce the net loss.

Another technique used to assign a value to life is the insurance approach; this gives a monetary worth to existence on the basis of individual life insurance decisions. This too has major drawbacks since bachelors with no dependants do not usually carry life insurance, yet this does not imply that they place no value on their lives. The court-decided compensation approach also has similar drawbacks because compensation for death or injury is usually calculated in relation to lost income. An alternative method is the implicit-value approach which assigns monetary worth to each life according to values implicit in past decisions affecting human mortality. Several estimates of the degree of expense that society has been willing to undergo to save a life have been made in France (Morlat, 1970) and the United States (Carlson, 1963). Values of a life developed using the implicit-value approach range from $9000 to $9 million, making it impossible to arrive at a realistic figure even if past decisions are accepted as having been correct (Henderson, 1975).

The safety committee might also consider applying a second group of techniques based on expressions of mortality risk in probabilistic terms (Linnerooth, 1976). The portfolio approach compares changes in mortality risk with the entire portfolio of risks assumed by society, while the utility or willingness-to-pay method places a value on risk reduction on the basis of the public's preferences or willingness to fund its mitigation.

Stress Safety Index

If the problem of assigning monetary values to lives saved and injuries prevented is seen as too cumbersome by the safety committee an alternative measure of safety might be an indicator based on the stress being experienced by residents of the community. The author has developed such an index elsewhere (Foster, 1976) and has shown that for the Developed World stress can be calculated from the formula

$$CS = 445a + 280b + cd$$

where

CS = the community stress experienced in the time period under review
a = number of deaths
b = number of serious injuries or illnesses
c = stress resulting from damage to the infrastructure (see Table 2.4)
d = population affected by the event

Although this technique has not yet been used to establish the

Table 2.4. Infrastructural Stress Values

Event intensity	Designation	Characteristics	Stress value
I	Very minor	Instrumental.	0
II	Minor	Noticed only by sensitive people.	2
III	Significant	Noticed by most people including those indoors.	5
IV	Moderate	Everyone fully aware of event. Some inconvenience experienced, including transportation delays.	10
V	Rather pronounced	Widespread sorrow. Everyone greatly inconvenienced, normal routines disrupted. Minor damage to fittings and unstable objects. Some crop damage.	17
VI	Pronounced	Many people disturbed and some frightened. Minor damage to old or poorly constructed buildings. Transportation halted completely. Extensive crop damage.	25
VII	Very pronounced	Everyone disturbed, many frightened. Event remembered clearly for many years. Considerable damage to poorly built structures. Crops destroyed. High livestock losses. Most people suffer financial losses.	65

beneficial effects of a community safety program, it has proved valuable in quantifying the stress associated with individual events. For example, the April 3, 1974, Xenia, Ohio, tornado caused over 6,300,000 stress units while the June 1972, Rapid City, South Dakota, floods were responsible for more than 6,600,000 (Foster, 1976).

If this formula were applied to illustrate changes in community safety, then stress associated with deaths and injuries could be determined from mortality and morbidity records, while that resulting from damage to the community's infrastructure would need to be calculated for each individual adverse event and summed (Table 2.4). An effective community safety plan should lead to a gradual reduction in the stress derived from the application of this index. If this does not occur, individual strategies, safety programs, and even the total plan will require revision.

Table 2.4. (*continued*)

Event intensity	Designation	Characteristics	Stress value
VIII	Destructive	Many injured. Some panic. Numerous normal buildings severely damaged. Heavy loss of livestock.	80
IX	Very destructive	Widespread initial disorganization. Area evacuated or left by refugees. Fatalities common. Routeways blocked. Agriculture adversely affected for many years.	100
X	Disastrous	Many fatalities. Masonry and frame structures collapse. Hazard-proofed buildings suffer considerable damage. Massive rebuilding necessary.	145
XI	Very disastrous	Major international media coverage. Worldwide appeals for aid. Majority of population killed or injured. Wide range of buildings destroyed. Agriculture may *never* be reestablished.	180
XII	Catastrophic	Future textbook example. All facilities completely destroyed, often little sign of wreckage, surface elevation may be altered. Site often abandoned. Rare survivors become lifelong curiosities.	200

References

Barker, M. L. 1977. *Specialists and Air Pollution: Occupations and Preoccupations*. Western Geographical Series, 14. University of Victoria, Victoria, B.C., 170 pp.

Behr, P. 1978. Controlling chemical hazards. *Environment,* **20(6)**:25–29.

Blume, J. A. 1978. Predicting Natural Hazards—State of the Art. *ASCE-ICE-CSCE 1978 Joint Conference on Predicting and Designing for Natural and Man-Made Hazards*. American Society of Civil Engineers, New York, pp. 1–10.

Bromenshenk, J. J. 1978. Yet another job for busy bees. *The Sciences,* **18(6)**:12–15.

Carlson, J. W. 1963. *Evaluation of Life Saving*. Unpublished Ph.D. dissertation, Harvard University, Boston.

Chapman, W. 1979. 'Harmless' drug takes a terrible toll. *Victoria Daily Times,* March 21, p. 40.

Christensen, H. (Ed.). 1976. *Suspected Carcinogens: A Subfile of the NIOSH Registry of Toxic Effects of Chemical Substances*. (2nd ed.). National Institute for Occupational Safety and Health, Cincinatti, Ohio.

Council for Science and Society. 1977. *The Acceptability of Risks*. Barry Rose, London, 104 pp.

Devons, E. 1961. *Essays in Economics*. Allen and Unwin, London.

Dublin, L. I., and A. J. Lotka. 1940. *The Money Value of a Man*. Ronald Press, New York.

Fischhoff, B., C. Hohenemser, R. E. Kasperson, and R. W. Kates. 1978. Handling hazards. *Environment,* **20(7)**:16–37.

Fisher, I. 1909. *Report on National Vitality*. Bulletin 3 of the Committee of One Hundred on National Health, U.S. Government Printing Office, Washington, D.C.

Foster, H. D. 1976. Assessing disaster magnitude: A social science approach. *The Professional Geographer,* XXVIII **(3)**:241–247.

Fraser, C. 1966. *The Avalanche Enigma*. John Murray, London. 301 pp.

Fromm, G. 1971. Civil aviation expenditures. *In:* R. Dorfman (Ed.), *Measuring Benefits of Government Investment*. Brookings Institution, Washington, D.C.

Henderson, M. 1975. The value of human life. *Search,* **6**:19–23.

Hoover, R., and J. F. Fraumeni. 1975. Cancer mortality in U.S. counties with chemical industries. *Environmental Research,* **9**:196–207.

Howe, G. M. 1970. *National Atlas of Disease Mortality in the United Kingdom*. (2nd ed.). Thomas Nelson, London.

Howe, G. M. 1979. Death in London. *The Geographical Magazine,* **LI(4)**:284–289.

Hutchison, G., and D. Wallace. 1977. *Grassy Narrows*. Van Nostrand Reinhold, Toronto.

Kitagawa, E. M., and P. M. Hauser. 1973. *Differential Mortality in The United States: A Study in Socioeconomic Epidemiology*. Harvard University Press, Cambridge, Mass. 255 pp.

Lamprecht, J. L., and S. J. Lamprecht. 1976. The geography of heart attack mortality rate in the OECD countries and its relationship to food consumption. *The Professional Geographer,* **XXVIII(2):**178–180.

Lawless, E. 1974. *Technology and Social Shock—100 Cases of Public Concern over Technology.* Midwest Research Institute, Kansas City, Missouri.

Linnerooth, J. 1976. Methods for evaluating mortality risk. *Futures,* **8(4):**293–304.

Lowrance, W. W. 1976. *Of Acceptable Risk.* William Kaufman, Los Altos, California.

Monmonier, M. S. 1974. Maximizing the information content of maps of spatial-temporal disease distributions. *The Canadian Cartographer,* **11(2):**160–166.

Morlat, G. 1970. *Un modèle pour Certaines Décision Médicales.* Cahiers du Seminaire d'Econométrie, Centre National de la Recherche Scientifique, France.

Murray, M. A. 1967. Geography of death in the United States and the United Kingdom. *Annals, Association of American Geographers,* **57(2):**301–314.

Otway, H. J. 1975. *Risk Assessment.* The Joint IAEA/IIASA Research Project, Vienna.

Puget Sound Council of Governments. 1975. *Regional Disaster Mitigation Plan for the Central Puget Sound Region.* 2 volumes. 106 pp and 208 pp.

Rabinowitch, E. 1972. Living dangerously in the age of science. *Bulletin of the Atomic Scientists,* **28:**5–8.

Reynolds, D. J. 1956. The cost of road accidents. *Journal of the Royal Statistical Society,* **119:**393–408.

Ridker, R. C. 1970. *Economic Costs of Air Pollution.* Praeger, New York.

Rodin, J. 1978. Predicting man-made hazards—state of the art. *ASCE-ICE-CSCE 1978 Joint Conference on Predicting and Designing for Natural and Man-made Hazards.* American Society of Civil Engineers, New York, pp. 21–31.

Schiel, J. B., and A. J. Wepfer. 1976. Distributional aspects of endemic goiter in the United States. *Economic Geography,* **52(2):**116–126.

Schoop, E. J. 1969. Development pressures after the earthquake. *In:* R. A. Olson and M. M. Wallace (Eds.), *Geologic Hazards and Public Problems.* Region Seven, Office of Emergency Preparedness, Santa Rosa, California, pp. 229–232.

Sewell, W. R. D. 1971. Environmental perceptions and attitudes of engineers and public health officials. *Environment and Behavior,* **3:**23–60.

Sewell, W. R. D., and H. D. Foster. 1976. Environmental risk: Management strategies in the Developing World. *Environmental Management,* **1(1):**49–59.

Shabecoff, P. 1978. Carter acts to soften dust law. *New York Times,* June 7.

Stamp. L. D. 1965. *The Geography of Life and Death.* Cornell University Press, Ithaca, New York.

Starr, C. 1969. Social benefit versus technological risk. *Science,* **165:**1332–1338.

The Sunday Times Insight Team. 1978. *Suffer the Children: The Story of Thalidomide.* Collins, London.

Theriault, G. 1977. Quoted in Birth defects blamed on gas. *The Victoria Times,* October 14, p. 13.

U.S. Department of Health, Education and Welfare, Public Health Service. 1974. *Health Characteristics by Geographic Region, Large Metropolitan Areas,*

and Other Places of Residence, United States 1969–70. Health Resources Administration, National Center for Health Statistics, 56 pp.
U.S. Environmental Protection Agency. 1977. *Toxic Substances Control Act (TSCA), PL-94-469: Candidate List of Chemical Substances,* 1–3. Environmental Protection Agency, Office of Toxic Substances, Washington, D.C.
Weidenbaum, M. 1978. *Impacts of Government Regulations.* Working paper No. 32, Center for the Study of American Business, George Washington University, Washington, D.C.
Weinberg, A. M. 1972. Science and trans-science. *Minerva,* **10:**209–222.

3

Development and the Spatial Distribution of Risk

No matter how much refined calculation has gone into the risk assessment, the scientific data are inevitably coarse, and estimates of consequences highly inexact. In the last resort the location of the practical limit of "acceptability" will be by fiat, based on personal judgements of those responsible for the decision.

Council for Science and
Society (1977)

Hazard Microzonation

The earth's surface is an intricate risk mosaic. The description and appreciation of this fact must necessarily form an essential ingredient in any rational attempt to satisfy community safety goals (Foster, 1975). Despite the obvious significance of spatial variations in the hazardousness of place, until quite recently, relatively few attempts have been made to quantify risk at the local level (Hewitt and Burton, 1971). This is not to imply that such differences in the occurrence of catastrophe have escaped widespread recognition. On the contrary, the extremely selective nature of much of the destruction has been recognized widely for many years. Snow avalanches, for example, may single out a house or a group of dwellings for ruination while neighboring structures are untouched, as occurred at Airolo, Switzerland, during February 1951 (Fraser, 1966). Earthquakes commonly selectively destroy, in a manner which to the cursory observer may appear quite random (Hodgson, 1964). The chaos caused by river flooding, seiches, storm surges, and tsunamis is also highly discriminatory (Office of Emergency Preparedness, 1972). The March 27, 1964, Alaskan earth movements, for instance, generated a series of seismic sea waves that left a trail of wreckage along the Pacific coast of North America, as interesting in its spatial variation as in its magnitude (Hansen and Eckel, 1966). Volcanic phenomena, such as the glowing avalanche of volcanic ash, debris, and poisonous gases which on 8 May 1902, was responsible for 30,000 fatalities in St. Pierre, Martinique, are also selective in their destruction (Macdonald, 1972). Spatial variations also occur in risks from hurricanes, tornadoes, fires, locusts, desert dust storms, dessication cracks (caused by the drying out of old lake silts and clays), surface collapses (the result of underground failure of the roofs of limestone caverns), soil burns (the ignition of upper organic horizons), and a wide variety of industrial and biological processes.

The effect of such spatial variation in destruction was exemplified by the devastation of the Los Alfaques tourist camp, near San Carlos de la Rápita in Spain, by the explosion of a propylene gas tanker (Harriss, 1979). Over 200 fatalities occured but their distribution was very uneven: "at one table three people had been calcined as they ate, while nearby another table was still set with flowers and a neat white cloth." Similarly, "beside a house trailer that had burned to the hubs hung a cage containing a bright yellow canary . . . its feathers only slightly singed."

Although traditionally the stimulus for belief in the supernatural, such selective decimation reflects differences in the distribution of risk rather than any plan of divine retribution. The total risk at any point is the result of a combination of all hazards and is measured by probability (Hewitt,

Table 3.1. Peacetime Disasters: Ratings of Community Disaster Probability

Name of Community ________ Date ______
Name of Respondent ________ Position ________

1. How would you rate the probability of the following events in your community, within this coming decade? Please rate them in terms of the following 6-point scale by circling the appropriate number.

0-Not applicable to my community
1-Not probable
2-Low probability
3-Moderate probability
4-High probability
5-Nearly certain

AVALANCHE	0 1 2 3 4 5
BLIZZARD OR MASSIVE SNOWSTORM	0 1 2 3 4 5
BOMB THREATS	0 1 2 3 4 5
CHEMICAL CONTAMINATION OR SPILL	0 1 2 3 4 5
CIVIL DISTURBANCE OR RIOT	0 1 2 3 4 5
DAM BREAK	0 1 2 3 4 5
DROUGHT	0 1 2 3 4 5
EARTHQUAKE	0 1 2 3 4 5
ELECTRIC POWER BLACKOUT	0 1 2 3 4 5
EPIDEMIC	0 1 2 3 4 5
FIRE	0 1 2 3 4 5
FLASH FLOOD	0 1 2 3 4 5
FOREST OR BRUSH FIRE	0 1 2 3 4 5
FREEZING ICE STORM	0 1 2 3 4 5
HOSTAGE INCIDENT	0 1 2 3 4 5

1970). It is inevitably a state of the world. Ignorance concerning this diversity of risk leads to uncertainty, errors in decision making, and frequently to avoidable disaster (Kates, 1970). There is clearly a need for hazard microzoning, the production of large scale maps depicting variations in degrees of risk, so that these can be given full measure in any disaster planning process. In this way high risk areas can be avoided or used for low intensity development only. It is with the production of such maps and the safety plan coordinator's role in ensuring their incorporation into normal regional and city planning that the remainder of this chapter is concerned.

Before any attempt can be made to quantify risk, the hazards threatening a society must first be identified. For this reason every community must carefully evaluate the types of risks to which it is subjected and rank these in the order of the threat to life and property that they pose. The British Columbia Provincial Emergency Programme has developed a simple probability table for this purpose (Table 3.1). It is reproduced in modified form here and should be completed by the safety plan coordina-

Table 3.1. (*continued*)

HURRICANE	0 1 2 3 4 5
LOST PERSONS	0 1 2 3 4 5
MAJOR FROST AND FREEZE	0 1 2 3 4 5
MAJOR GAS MAIN BREAK	0 1 2 3 4 5
MAJOR HAIL STORM	0 1 2 3 4 5
MAJOR INDUSTRIAL ACCIDENT	0 1 2 3 4 5
MAJOR ROAD ACCIDENTS	0 1 2 3 4 5
MAJOR WATER MAIN BREAK	0 1 2 3 4 5
MINE DISASTER	0 1 2 3 4 5
MUD OR LANDSLIDE	0 1 2 3 4 5
OIL SPILL	0 1 2 3 4 5
PIPELINE EXPLOSION	0 1 2 3 4 5
PLANE CRASH IN COMMUNITY	0 1 2 3 4 5
RADIOLOGICAL ACCIDENTS	0 1 2 3 4 5
RIVER FLOOD	0 1 2 3 4 5
SEVERE FOG EPISODE	0 1 2 3 4 5
SHIP DISASTER IN HARBOUR OR NEARBY COAST	0 1 2 3 4 5
SMALL BOATS, LOST OR ACCIDENTS	0 1 2 3 4 5
SMOG EPISODE	0 1 2 3 4 5
SUDDEN WASTE DISPOSAL PROBLEM	0 1 2 3 4 5
RAILWAY ACCIDENTS	0 1 2 3 4 5
TORNADO	0 1 2 3 4 5
TSUNAMI	0 1 2 3 4 5
VOLCANIC ERUPTION OR FALLOUT	0 1 2 3 4 5
WATER POLLUTION	0 1 2 3 4 5
WATER SHORTAGE	0 1 2 3 4 5
OTHER	0 1 2 3 4 5

Source: British Columbia Provincial Emergency Programme.

PLATE 4. On April 5, 1974 a tornado severely damaged one-half of Xenia, Ohio. Tornadoes are a major threat to many communities (American Red Cross photograph by Ted Carland).

tor and associated committee. It should also be widely circulated to obtain an overview of perceived risks in the community at large. In addition, a group of experts might also be asked to take part in a Delphi study of local hazards to aid in identifying their relative significance. The methodology required is described elsewhere in this volume. Naturally those damaging events that are most likely to occur and that have the greatest potential for devastation should be given priority in the production of risk maps.

Data Collection

Before risk levels can be effectively utilized to control development several tasks must be accomplished. First, differing hazard zones must be delineated. Second, loss potential of alternative combinations of structural designs and land use activity within these zones must then be established. Finally, standards must be set for unacceptable risk to life and property (Puget Sound Council of Governments, 1975). The first step in this process, the production of maps defining spatial variations in risk, requires data from a wide variety of sources. Ideally, direct observations specifically taken to determine the frequency, magnitude, and intensity with which a threatening event occurs are required to produce such microzonations. In the Developed World, for certain hazards such as floods, avalanches, forest fires, tornadoes, hurricanes, and air and water pollutants, there is generally a national or regional agency which has been collecting and processing such data for many years. Groups of industries, unified by a common interest, often maintain their own association which frequently has an information collecting function. Data from these and other sources is often freely available to local officials in either published or computerized form. The Smithsonian Institution (1971) has attempted to provide a complete listing of such information centers and it is advisable to consult their published directory. In this volume it is obviously impossible to provide such a comprehensive survey of data sources and one example must suffice. Some 200 Canadian cities are at risk from floods and several federal–provincial initiatives are underway to produce microzonations of the hazard. Where they are not yet available from either federal or provincial sources, the data generally exist to permit local planners themselves to determine spatial variations in risk from floods. The Water Survey of Canada, Inland Waters Directorate, Environment Canada, for example, publishes provincial discharge surveys, such as its *Historical Streamflow Summary British Columbia to 1973* (1974). Volumes of this nature normally list among other data the extreme recorded discharge for the period of record. Other federal or regional reports may describe the flood history based upon such data, as has been the case for the New Brunswick-Gaspé region (Collier and Nix, 1967). In the United States, similar flood information is freely available from various federal agencies including the Geological Survey, the Army Corps of Engineers and the Department of Agriculture (U.S. Office of Emergency Preparedness, 1972). Many flood microzonations have already been completed, including those for the Chicago area described by Sheaffer, Ellis, and Spieker (1969).

Assuming the levels of risk normally considered to be acceptable and the frequency of major catastrophes it would take thousands of years to establish the risk level of some hazards solely by direct observation. Clearly, planners cannot simply rely on experience alone to evaluate disaster risk (Steinbrugge and Bush, 1965). Communities cannot wait for several large floods, fires or nuclear plant failures before taking action. The data base consisting of scientifically collected information must, therefore, be supplemented by other less accurate but nevertheless essential sources. Fortunately many of these are available in most areas. These include academic, business, and government scientific papers, maps, and records. Contour maps, for example, allow slope to be calculated, a necessary consideration in the mapping of landslides and other mass movement phenomena. Many municipalities and private drilling companies maintain borehole records which include information on the location of wells, the depths at which changes in stratigraphy were encountered, and the heights of the water table. This type of data is extremely valuable in the development of microzonations for earthquakes and for predicting pollution risks from waste disposal sites (Wuorinen, 1976; Fenge, 1976). Many local governments also compile telephone, drain, and sewer profile records which show the nature of the urban substrate encountered when such underground facilities were installed. These can be used in flood, earthquake, subsidence, pollution and instability microzonations. Figure 3.1 is an earthquake risk map for the Victoria area based largely upon such records. Similarly, published accounts and maps of bedrock geology, soils, forest cover, and land use can be used as inputs into the microzonation process. These are usually available from sources such as national and regional forestry services and geological and soil surveys.

Historical visual information, films, photographs, and diagrams available from government and university archives, newspaper files, television studios, and private collections may also be of value. These visual records are commonly compiled immediately after disasters, since such events are particularly newsworthy. The information films and photographs contain often allows spatial differences in destruction, and hence the magnitude of impact of past events, to be established with considerable accuracy. Where damage may occur more slowly, as with coastal losses, comparisons of older ground or aerial photographs with those taken more recently at the same location permit estimates of the rate of erosion. In a study of the decline in the recreational potential of a provincial marine park on Sidney Island, British Columbia, caused by sandspit erosion, letters were sent to the editors of all local newspapers. This correspondence successfully appealed to readers for access to old photographs of the spit. Several responded to the request and approximate rates of erosion were established by the author from the material provided (Foster and Norie, 1978). Aerial or satellite photographs taken during disaster are particularly useful in establishing risk zones. In some cases, as with U.S. Geological Survey maps of the flood risk at Rapid City, they are actually incorporated into the final microzonation.

Private and public records of past and present industrial activity are also pertinent. An analysis of business locations from old city or telephone directories, for instance, can help in locating premises where hazardous materials, including radioactive substances, may have been used without adequate safeguards. Former waste disposal sites may also be located in this manner and their safety assessed. Three recent cases illustrate the need for such caution. In February 1976, the Canadian federal government announced that it was aware of 109 certain or suspected locations which had been contaminated by radioactive materials. In Port Hope, Ontario, probably the worst case, schools were closed and residences evacuated because of this hazard. Genetic damage to future generations has been predicted for the area's resi-

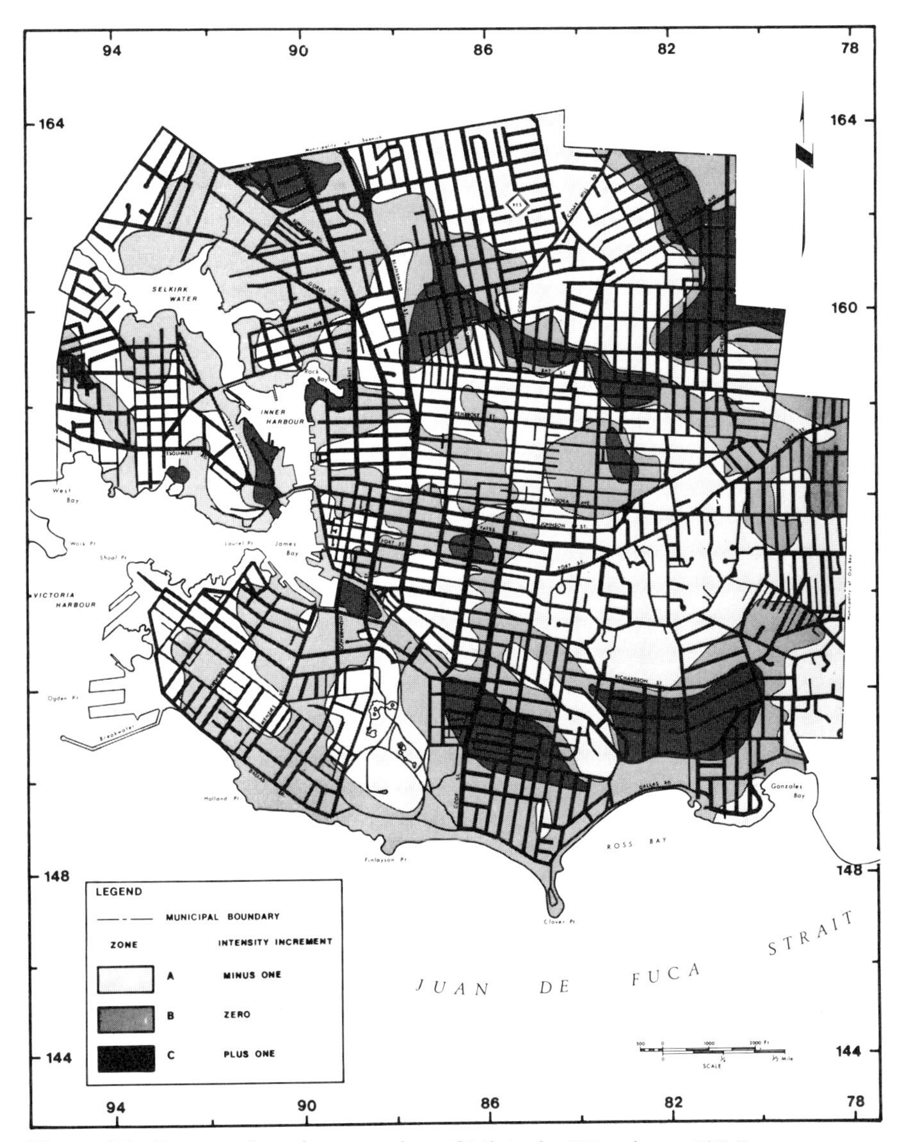

Figure 3.1. Earthquake microzonation of Victoria (Wuorinen, 1976).

dents. In Warsaw, Poland, two tons of cyanide were found in rotting barrels in an abandoned basement workshop, where formerly it had been used in a galvanizing process. Officials estimated that had it leaked into the water supply it could have killed the city's entire population.

One of the most dramatic examples of the risk posed by waste disposal is that of the Love Canal, Niagara Falls, New York. For approximately 10 years the Hooker Chemicals and Plastics Corporation used the canal as a dumping ground for more than 80 chemicals. The site was then covered with earth and the property turned over to the city and school board. A school was built on the dump and homes were erected on streets bordering the former canal. New York State declared a health emergency and staged an evacuation of the neighborhood in the summer of 1978 as chemicals oozed to the surface. Studies indicated extremely high rates of miscarriage and birth defects in the area. Chemicals also appear to have migrated along adjacent streambeds and into the basements of many homes, increasing the incidence of a wide variety of diseases among residents. It has been estimated that, in the United States alone, there are perhaps as many as 1000 similar abandoned chemical disposal sites.

In addition to checking the locations of former industrial plants and those still in production, ports and coastal settlements would be well advised to examine the cargo records of any wrecks in their vicinity and to be cognizant of the type and magnitude of hazardous substances traveling their local waters or indeed their road systems.

The records and logs of insurance agencies, police, fire, hospitals, and disaster-related agencies or societies, such as the Red Cross, are further sources of useful information. These often provide detailed information on actual losses from which locations and magnitude of disaster can be estimated. Newspaper and magazine articles can also be of value as can diaries kept by local inhabitants. For instance, it appears from the Rev. F. Consag's records that the last time Tres Virgines volcano in the Gulf of California erupted was in 1746 (Ives, 1962). A description of a 1788 tsunami generated in the Aleutian Islands area has been found among documents of a Russian–American company operating there during the period 1799–1867. This appears to have been one of the most catastrophic seismic sea waves ever recorded, comparable to that generated by the 1964 Alaskan earth-movements (Solov'yev, 1968). Statistical analysis of health records or death certificates may also allow hazards to be identified and microzones showing differences in related risks to be prepared. Lung cancer rates in the Los Angeles and San Francisco areas are more than double the United States average, with San Mateo County on the fringes of the latter city being one of the worst areas. Chicago, in contrast, has a lung cancer rate of less than nine per 100,000 population, while San Mateo County's rate was 51.7 in 1975. The national U.S. average is about 19 per 100,000 population. The major cause of the spatial differences in the risk of such cancers appears to be air pollution, those living in close proximity to oil refineries and other heavy industry being particularly at risk.

Major damaging events may not have occurred locally during the period of recordtaking. This precludes direct observation and instrumentation as the basis for microzonation. In such circumstances the possible magnitude and distribution of disaster agents may be predicted by the application of mathematical formulas. These have generally been developed from experience of the hazard gained elsewhere and the validity of their use usually depends on a similarity of physical characteristics. Commonly used formulas are those for predicting peak river discharges. Such flood formulas require information on the location and size of the drainage basin involved. Numerous flood formulas exist and the validity of their use depends on correct selection. Hydrometeorological methods can also be used to predict maximum possible floods, these models are based on assuming the worst combination of weather and hydrological conditions. Examples of their use include the design of the dams of the Australian Snowy Mountains scheme and in calculating the needed spillway capacity of the Mangla Dam in West Pakistan (Binnie and Mansell-Moullin, 1966). Once possible peak discharge has been predicted or floods of lesser magnitude simulated, a knowledge of anticipated velocities and the relief of the area allows predictions of the scale and intensity of flooding. This information can then be used to produce a microzonation.

Another approach to prediction of the spatial distribution of disaster has been the use of statistical methods. Gumbel (1958) has been prominent in developing statistical distributions that describe the occurrence of extreme events. He suggests that the recurrence intervals of exceptionally large phenomena bear consistent relationships to their magnitude, expressed in either arithmetic or logarithmic terms. If this is so then say 50 years of data can be used to extrapolate to determine the once-in-a-1000-years event. This and other magnitude information might then be used to microzone (Chorley and Haggett, 1967). It should be noted that this concept has been criticized on the grounds that the confidence limits are generally so large that the resulting estimates are largely meaningless (Wiesner, 1964).

Some predictive models can be used to produce microzonations directly. The universal soil loss equation, largely developed by the Runoff and Soil-Loss Data Center of the U.S. Agricultural Research Service, located at Purdue University, is such a tool. Its use permits the direct production of a microzonation map showing differences in rates of soil erosion and probable associated sedimentation problems. This predictive equation is as follows:

$$A = RKLSCP$$

where A = soil loss in tons per acre (normally calculated annually), R = rainfall factor, K = soil erodability factor, L = slope-length factor, S = slope-gradient factor, C = cropping management factor, and P = erosion control practice factor. Several of these factors, such as R, L, and S, can be obtained from tables, while others can be derived from the same sources after field investigation (Beasley, 1972). Numerous other predictive models exist which permit the calculations necessary to produce microzonations for a wide range of further natural and man-made hazards (Office of Emergency Preparedness, 1972; Puget Sound

Council of Governments, 1975a, 1975b). They include, for example, numerical simulations of snow avalanche flow (Lang, Dawson, and Martinelli, 1979).

Almost every area was settled for many years, perhaps even many centuries, before environmental and social data were scientifically collected. In regions where monitoring networks are relatively new, their information base may be extended by or at least compared against the oral history of the district. Two major sources exist. There are eye-witness accounts of past events as remembered by long-time residents and legends passed down by word of mouth. Many archives and some universities have oral history departments, the members of which record or take notes at meetings with pioneers or older members of the community at which these elderly residents describe memorable events. Such information may be of value in microzonation since disasters, particularly those in which the person interviewed was a victim or nearly so, are likely to be remembered with some clarity. If no such data bank exists, appeals on the radio, television, or through local newspapers may bring likely respondents forward for interview.

Legends of catastrophe are commonplace since such events may form watersheds in the history of a society (Vitaliano, 1973). Although these may be very distorted accounts of actual events they can be of great value in identifying threats with long-term return periods. Vitaliano, in her book *Legends of the Earth*, describes hundreds of such oral records and provides an ideal reference source. She points out, for example, that the Bronze Age eruption of Santorin in the Mediterranean has been suggested as the direct or indirect cause of legends which include those of Deakalion's deluge (Galanopoulos, 1960), the sudden collapse of the Minoan civilization on Crete and the concomitant rise of the Mycenean civilization on the Greek mainland (Marinatos, 1939); the myth of Atlantis (Galanopoulos, 1960), for the biblical plagues of Egypt (Bennett, 1963), the parting of the waters of the Red Sea (Galanopoulos, 1964), the myths of Phaëthon (Galanopoulos, 1969) and Icarus (Brumbaugh,1970), and parts of the Theseus legend (Renault, 1962). Natives in North and South America, Africa, and elsewhere have preserved an abundance of comparable legends, many of which describe the location and scale of damage associated with catastrophic events (Vitaliano, 1973). These can be used as a starting point for geological, geomorphological, and stratigraphical studies to identify zones of risk.

Certain catastrophic events take place so rarely in any one location that there may be no locally available data on which to predict risk and produce a microzonation for the hazard involved. This should not be misinterpreted to mean that there is no danger. In such circumstances, microzonation may be based on argument by analogy. Consider, for example, the risk of a catastrophic eruption in the Cascades, perhaps equalling or excelling that of Santorin volcano in Greece or Krakatoa, Indonesia (Macdonald, 1972). The major damage, apart from localized destruction of the total infrastructure, would result from volcanic ash. This can travel hundreds of miles before it settles and causes extensive losses to buildings, crops, animals, and communications systems. It is extremely difficult to predict the probability of such a rare event. It cannot, however, be discounted since a comparable eruption took place at Mount Mazama, Oregon, approximately 6600 years ago. This pumic eruption removed several cubic miles of materials to create what is now Crater Lake and blanketed an area of several hundreds of thousands of square miles in northwestern Canada and the United States with ash. In Figure 3.2 the 6-inch fallout area for pumice is shown superimposed on other volcanoes in the Cascade Range to give an idea of the areas likely to be most seriously affected if a similar type and scale of eruption occurred at any one of them (Crandell and Waldron, 1969).

Similar argument by analogy in the production of microzonation maps may be used with man-made hazards. In their attempt to determine total risk in London, Ontario, Hewitt and Burton (1971) sent questionnaires to 100 North American cities of relatively similar size. These solicited information on the occurrence of both natural and man-made disasters which

PLATE 5. The bursting of the Teton Dam, June 1976, destroyed or damaged more than 4000 dwellings, including this one in Blackfoot, Idaho. Dams are a common man-made hazard that must be given cognizance in the microzonation process.(American Red Cross photograph by Gibson).

the intensity of drought problems (U.S. National Water Commission, 1973). As with many other hazards, a map of distribution is essentially a microzonation illustrating regional differences in loss potential (U.S. Department of Agriculture, 1973).

Disease or damage to vegetation and animal life may be an indication of the adverse impact of other hazards and can therefore be used indirectly as indicators of their distribution. Such information, for example, is commonly a key source in the production of avalanche microzonations. In Rogers Pass, British Columbia, used by the TransCanada Highway to cross the Selkirk Mountain Range, vegetation has been used as an indicator of risk (Schaerer, 1972). Those trees selected for study were fully exposed to avalanches and not protected by boulders or irregularities in the terrain. It was found that where alder and willow brush occur on the outrun zones, large avalanches take place frequently, usually once a year. Where there are few large trees and no branches on the side facing the avalanche zone below 26 ft above the ground, avalanches occur once every one to three years. These are smaller in size and produce mainly windblast and deposit snow not deeper than 3.3 ft. Where there are no large trees and no deadwood from such vegetation but spruce trees 11.5–15 ft high and hemlock 5–6.5 ft tall with damaged branches and bark, large avalanches occur once every three to ten years, depositing deep snow. In areas where there are broken large trees and a dense growth of smaller trees, large avalanches take place infrequently, not more often than once a decade and probably not even once in 20 years. Using this and other data it was possible to microzone for avalanche hazard in Rogers Pass and plan control mechanisms to protect the major highway it carries. Figure 3.3 provides an illustration of the use of irregularities of tree growth and the absence of lower limbs to establish the frequency and magnitude of flooding, in this case on the Potomac River near Chain Bridge (Pitty, 1971). It should be noted that vegetation may also be used to establish when and how often damaging events have occurred through the use of such techniques as radiocarbon dating, lichenometry, the measurement of growth rates of lichen on devastated or newly deposited surfaces. and dendrochronology, the counting of tree rings (Pitty, 1971). Madole (1974) used many of these techniques to delimit risk from avalanches and other hazards in Colorado.

Vegetation damage may also be an indication of man-made disaster agents, reflecting airborne, soil, or water-carried pollutants or pests. Downwind of the Alcan Aluminum plant in Kitimat, British Columbia, there is a wedge of dead or dying coniferous forest many miles in length. Workers at this factory insist that this is caused by chemical poisoning which is also responsible for many illnesses among them, while management argues that it is due to forest pests attacking the trees in question. In either case the health of the forest could be used to produce a microzonation of the hazard involved.

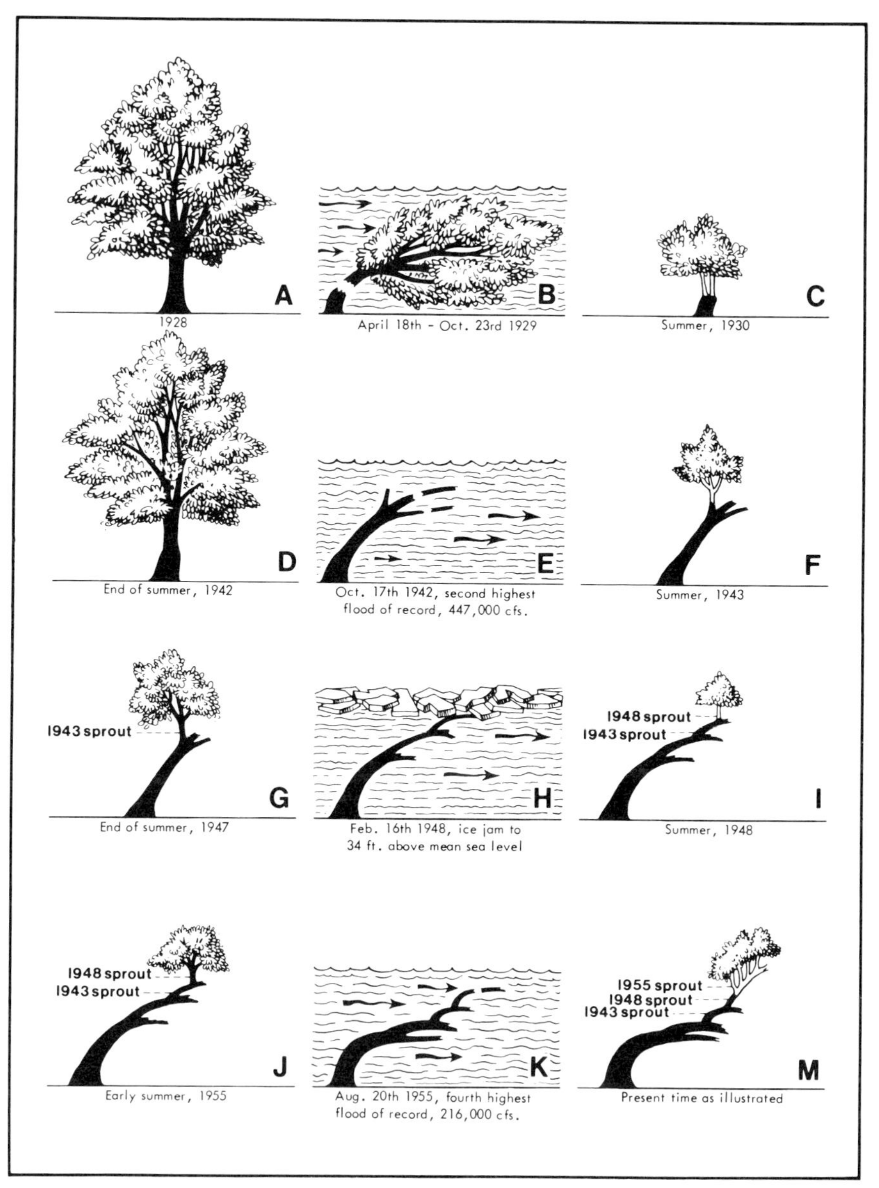

Figure 3.3. Sprouts of four ages on flood-damaged ash tree, Potomac River (after Sigafoos, 1964, cited in Pitty, A., 1971, *Introduction to Geomorphology*, Methuen: London).

The landscape itself carries the scars of former periods of intense geophysical and meteorological violence. Many landforms and the deposits of which they are composed provide evidence of destructive and constructive processes which, if repeated after settlement, would be termed disaster agents. The interpretation of these geomorphological features and their associated deposits can provide information of spatial differences in risk, often expanding the instrumental record by hundreds or perhaps thousands of years. This approach is particularly useful in determining the frequency and intensity of very rare, yet extremely damaging events, like major volcanic eruptions and flash floods. Such studies rest upon the identification of particular landforms and deposits. These are then subjected to morphological examination to determine the magnitude and direction of movement of the agents which created them. Deposits can be analyzed to determine such variables as the velocity of flow or rates of settling, distance, and direction of travel and age (Pitty, 1971).

Volcanic hazards in the U.S. Cascade Range have been studied using such geological and geomorphological techniques (Crandell and Waldron, 1969). More detailed examinations have also focused upon individual volcanoes such as Mount Rainier, Washington (Crandell and Mollineaux, 1967). The latter study determined that damaging debris flows could be expected from Mount Rainier once in 10 to 100 years. These are by far the greatest hazard because of their high frequency, possible large size, and concentration along valley floors. The area suffered from both the Osceola mudflow which occurred 5000 years ago, and the Electron mudflow some 500 years in age. Indeed, within the last 10,000 years Mount Rainier's volcanic activity has resulted in at least 55 large mudflows, several hot avalanches of rock debris, 12 volcanic ash eruptions, and one or more lava flows. The Osceola mudflow, one of the largest lahars, contained 2.5 billion cubic yards of material and covered 125 square miles in the Puget Sound lowland. This area now has a population in excess of 30,000 and includes the sites of the towns of Enumclaw, Buckley, Auburn, and Sumner (Crandell and Waldron, 1969). These events have not occurred at random around the flanks of the volcano, and some valley bottoms, especially that of the White River, could be zoned extremely high risk on the basis of the geological record. The most detailed volcanic microzonation maps known to the author are those that have been produced for several Javanese volcanoes, including Mt. Merapi and Mt. Kelut, drawn at a scale of 1:50,000. An example is provided in Figure 3.4. These clearly demonstrate that most high risk sites are those in, or down slope from valleys, gullies, and other depressions which act as the paths of least resistance for glowing avalanches, volcanic mudflows, and lava (Geological Surveys of Indonesia, 1974).

Because of the short time span for which records have been kept, the risk of rare yet devastating catastrophic floods is very difficult to evaluate using conventional flood frequency analysis. In such circumstances the stratigraphical record of the Holocene, the past 10,000 years, may be a better guide to the distribution of risk. Numerous techniques can be used to interpret this record, including tree ring analysis, radiocarbon dating of eroded or deposited sediments, and buried landform analysis (Costa,

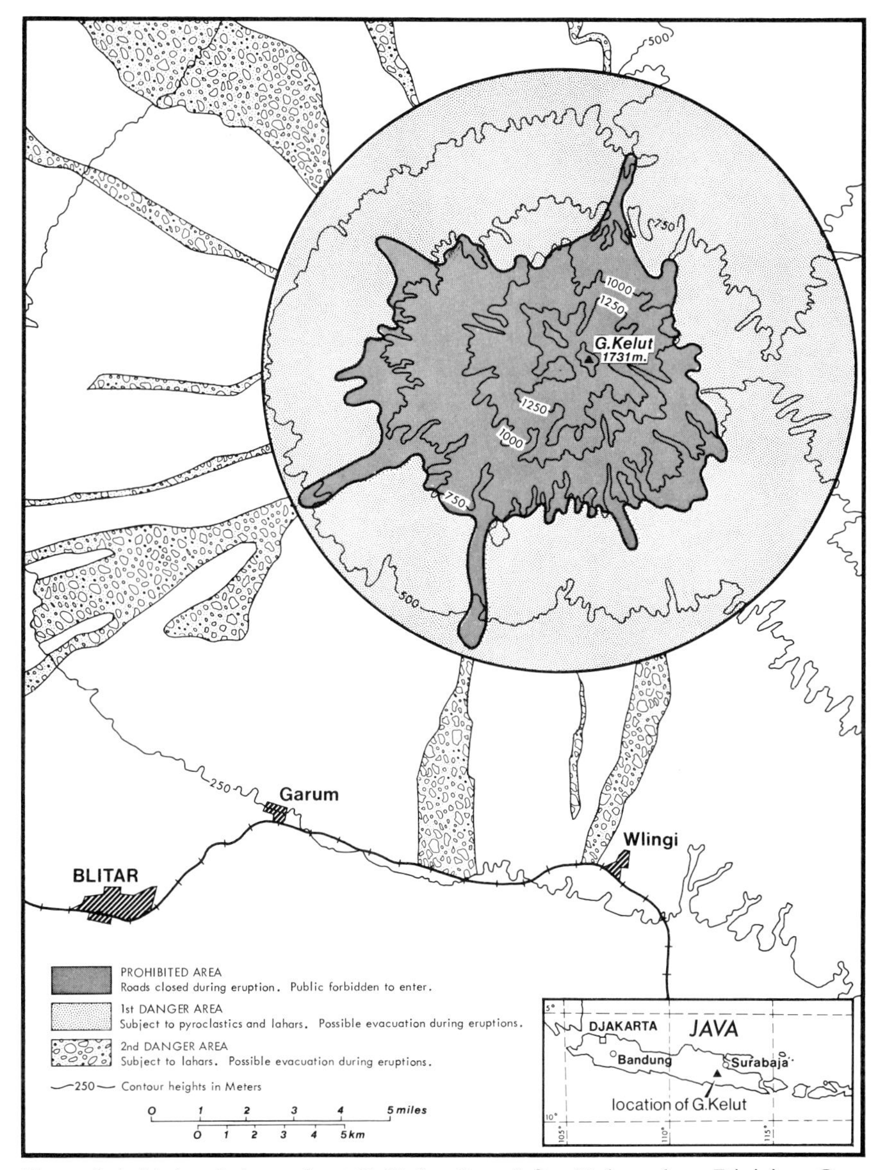

Figure 3.4. Volcanic hazards at G. Kelut, Java (after Volcanology Division, Geological Survey of Indonesia).

1978). Using one or more of these techniques, Costa was able to establish the return period of several recent catastrophic floods in the United States: 5000 years for the Big Thompson River event of 1976; 100 years for the 1964 Klamath River flood; 2100 for the 1972 Western Run, Maryland, inundation, and 400 years for the 1972 devastation of Elm Creek, Texas. On the basis of this type of information the hydrological record can be extended and large areas zoned according to their risk from floods.

In addition to these hazard-oriented studies there are many thousands of meteorological, geomorphological, or geological publications written for other purposes. These have included a desire to establish the age or origin of particular landforms or to describe the locations of epicenters of earthquakes. This material can often be reworked from a microzonation viewpoint. Landslide and avalanche studies may be of value in this way as is research into the distribution of sand and gravel deposits for commercial purposes. In fact, most soil, geological, and geomorphological surveys, even when conducted for other reasons, can be extremely useful in the production of hazard maps.

Records of building damage are also of great value. Immediately after a disaster, associated structural damage and resulting casualties should be assessed on a geographical basis. Such a survey should be undertaken as quickly as possible, so that the evidence is not obliterated by cleanup operations and repairs. This process is greatly facilitated if a photographic file showing the state of repair of a cross section of buildings in the area has been established before the disaster agent strikes. This allows recent damage to be distinguished from preexisting problems and promotes a realistic assessment of structural response to disaster agent impact. The compilation of such a photographic file can be combined with examinations of older buildings for evidence of former disaster damage, such as earthquake or subsidence-induced cracking. Such information can be used to extend the modern data base.

In the same way, archaeological excavations of ancient buildings or the state of repair of ruins may provide information on the cause and date of their destruction. Such surveys can therefore be used to help determine the magnitude of past disaster agents and spatial variations in their intensity of impact and frequency of occurrence. This information is valuable for many reasons, for use in the revision of building codes, for improving construction materials. simulating disasters, and in the producing of microzonations. (Texas Coastal and Marine Council, 1977; Sumi and Tsuchiya, 1976; Penning-Rowsell and Chatterton, 1977).

There are numerous examples of the use of such building surveys and archaeological excavations to determine disaster potential. On May 8, 1902, a glowing avalanche (nuée ardente) from the volcano, Mt. Pelée, struck St. Pierre, Martinique, killing its 30,000 inhabitants. Its impact knocked over and tore apart masonry walls 3 ft thick, uprooted large trees, swept 6-inch cannons from their mounts, and carried a 3-ton statue 40 ft from its base. This and other structural evidence makes it clear that the cloud must have been traveling at a velocity of some 100 miles/hour. Injuries to the dead were grotesque, bodies were intensely burned, often stripped of clothing by the blast, and skull sutures opened up. The injuries were such as would result from sudden heat, intense enough to form steam in tissues but not capable of raising clothing to kindling temperatures (Macdonald, 1972). Estimates therefore place the temperature at between 700°–1000°C.

Excavations on the volcanic island Santorin and elsewhere around the Mediterranean indicate that a major eruption which created an enormous tsunami took place approximately 1450 to 1480 B.C. Archaeological evidence indicates that this event, perhaps followed by numerous violent earthquakes, was largely responsible for destroying most of the coastal Minoan cities on Crete. On the island of Anafi, 15 miles east of Santorin, a layer of pumice 15 ft thick occurs at an elevation of 820 ft, showing a minimum wave height. A building on the island of Amnisos had been swept by large waves, pumice and sand being found in cavities within the structure and in its foundations. On the same island in the villa of the frescos, parts of the walls and corners of the room of that structure have collapsed in an unusual manner, the walls bulge outward, and large monoliths weighing several tons have been pushed out of position or are completely missing (Vitaliano, 1973). Such archaeological evidence allows estimates of the height, intensity, and age of the tsunami to be made with relative accuracy. Similarly, after the 1964 Alaskan tsunami, surveys of the structural damage it had caused allowed its intensity and spatial impacts to be established with relative accuracy (Tudor, 1964; Magoon, 1966).

A wide range of studies which included detailed damage-related tables are available to help estimate the magnitude and intensity of disaster agents from their resulting destruction (Whitman, 1973; Steinbrugge, 1970; Penning-Rowsell and Chatterton, 1977). Consider, for example, the damage caused to buildings by avalanches. This occurs in several different ways. First is the direct thrust, with maximum impact pressures of very large, fast-moving avalanches exceeding 100 tons/square meter (Salm, 1965). In addition, avalanches may exert upward and downward forces; they have lifted and moved large locomotives, road graders, and buildings (Perla and Martinelli, 1976). Field measurements indicate that upward and downward forces are about one-fourth to one-half the direct thrust. Finally, the airblast of rapidly moving avalanches with speeds in excess of 30 meters/second may exert pressures up to about 0.5 tons/meter, capable of destroying doors, windows, and poorly designed roofs. Table 3.2 shows the relationship between impact pressure and potential damage (Perla and Martinelli, 1976). It could be used to assist in the producing of microzonations in avalanche-prone areas. This is comparable to other information for a variety of disaster agents and is extremely useful in extending and amplifying the actual observed record.

Where there is no precedence there can be no experience. Risk related to potentially hazardous innovation is, as a result, the most difficult to quantify and to microzone. Much of the debate and legal confrontation between technocrats and enviromentalists centers on the quantification of such risks and disagreement over what constitutes an acceptable level of

Table 3.2. The Relationship between Avalanche Impact Pressure and Potential Damage

Impact pressure (t/m²)	Potential damage
0.1	Break windows
0.5	Push in doors
3.0	Destroy wood-framed structures
10.0	Uproot mature spruce
100.0	Moves reinforced concrete structures

Source: Perla and Martinelli (1976).

associated social vulnerability. Just how significant the risk factor in innovation may prove to be has been illustrated by the testing of nuclear weapons. In 1953 St. George, Utah, was blanketed by fallout from atomic bomb tests in Nevada. Lyon (1979) has since demonstrated that childhood leukemia doubled in southern Utah during the years of such atmospheric testing. So-called safety tests, to demonstrate that a nuclear chain reaction would not be touched off if a conventional bomb hit an atomic weapon, were also held in Nevada. As a result, plutonium contamination of many of the state's soils occurred, which may well result in continuing high cancer death rates (United Press International, 1979). Innovation, therefore, poses a major problem in defining risk and should be subjected to particularly stringent controls.

For many hazards, however, there is considerable data on past frequency and intensity of impact. Once this has been collected it can be presented as a series of maps. These are of value in forecasting disasters and in promoting the adoption of strategies to avoid them. Such information may require further analysis through the use of one or more predictive models, before this goal can be fully achieved. This analytical process is described in more detail in Chapter 4. Naturally, for many hazards data on past occurrence and impact may be limited, or even unavailable. Similarly, for some rare events the past may not be a sufficient guide to the future. Nevertheless, planning decisions are being made daily, based on incomplete information. For this reason, it is essential to use the data on risk that is available, even if it is not as detailed or as comprehensive as might have been wished.

Scale

Once information has been collected to aid in the delineation of risk, it becomes possible to map vulnerability at a variety of different scales. While those maps at a scale of smaller than 1:63,360 may prove useful as descriptive tools, gaming aids or political attention-attracting devices, they are of little use in actual planning. A variety of larger scales have been used for this purpose. The Tennessee Valley Authority and the United States Army Corps of Engineers, for example, have been mapping flood plains at scales of one or two inches to 1000 ft. It has been realized that even at the larger of these two scales, it would be difficult to determine the threat to individual buildings with any degree of precision. For this reason, the Tennessee Valley Authority is now promoting the use of hazard maps at a scale of one inch to 400 ft, as a minimum standard in urban areas (Office of Emergency Preparedness, 1972).

Despite the limitations implicit in more generalization, almost all hazard microzonations have been produced at a smaller scale than this. Juneau, Alaska, for instance, has been mapped to portray avalanche hazards at a scale of 1:40,000. This map recognizes three spatial variations in risk, delineated as zero, potential, and high hazard (White and Haas, 1975). This contrasts with use of a scale of 1:8,000 by Ives, Mears, Carrara, and Bovis (1972) when mapping Ophir, Colorado, for the same disaster agent. The latter map is sufficiently precise to permit building-by-building risk discrimination while the former is not. Numerous other microzonations have been produced at a wide variety of horizontal equivalents. These include tsunami innundation maps of Hawaii at scales of 1:63,360, and Tokyo's and Victoria's seismic risk at 1:50,000 and 1:12,000 respectively (Ohsaki, 1972 and Wuorinen, 1974).

The safety plan director and committee should pay considerable attention to the question of mapping scale for hazard microzonations. All that can really be suggested is that the scale selected should be large enough to make full use of the available data and to permit individual sites and structures to be identified if possible. It should not be so large, however, that it gives an invalid impression of precision in areas where the information does not substantiate such a position. The number of risk zones to be delineated should be determined in a similar manner.

Single Hazard–One Purpose Mapping

The simplest microzonations are one hazard–one purpose maps designed to show the probable areal extent of a single event or phenomemon. Of these maps, the least sophisticated are those based upon intelligent guesswork, as was the case of flood hazard maps prepared for Laona Township, Illinois:

> They tell me that a storm drainage plan involves engineering analysis, appraisal of land form, gradient, static and fluid capacity, outfall, watershed and time capacity factors, and determination of the amount of periodic rainfall. But this would take thousands of dollars, months if not years. . . . No, we in Laona Township were too conscious of nature and man's foolishness to wait without protection. Simple minimum standards had to be found and secured by zoning.

> So every stream and draw in Laona Township that drains water from more than 500 acres is labelled floodway. All future structures and fences, other than agricultural fences, must be at least 150 feet back from the centre line of the stream. . . . No building may be built with a floor elevation lower than 15 feet above the stream at its lowest point (White, 1964).

This decision to preserve a flood channel 300 ft wide was taken solely on the basis of local experience and common sense, bypassing any elaborate survey or calculations.

More commonly, one hazard–one purpose maps are produced using a more complete data base but still with a single objective in mind. They might be used, for example, to show the present distribution of *Coccidioidomycosis immitis,* the disease causing fungus found in some of the arid soils of the western hemisphere (Welsh, 1973) or the extent of the average distribution of air below a certain quality (Fosberg, 1973). Many such maps are required by fiat since they are needed to implement legislation or agency policy. Examples include microzonations showing areas of potential tsunami inundation in the San Francisco Bay region, compiled by the United States Geological Survey in cooperation with the Department of Housing and Urban Development in 1972. These risk maps assume a runup of 20 ft at Golden Gate and indicate the areal extent of the resulting flood (Ritter and Dupre, 1972). Similar maps showing the potential for tsunami inundation in Oahu have been prepared by the Hawaiian Civil Defense and McDonald, Shepard, and Cox (1947). These are illustrated in Figure 3.5.

Single Hazard–Multiple Purpose Mapping

Slightly more sophisticated are the single hazard–multiple purpose microzonations, designed to show the areal extent and intensity of impact of one disaster agent but on several possible occasions. Repeated experience has shown that earthquake damage can vary greatly, even within small areas. It is generally agreed that, constructional factors being equal, buildings on deep moist clays, sands, or particularly fill, especially if on or near an active fault will suffer the most damage during an earthquake. It is therefore quite feasible to microzone urban areas according to their seismic vulnerability. Such maps, first used for Tbilisi in the Soviet Union, are currently available for Sofia in Bulgaria; Bucharest, Galati, and Arges in Romania; and for several Turkish, Yugoslavian, and Soviet cities (Kárník, 1972). At least six Chilean cities, including Santiago, have been microzoned, as have Boston, Seattle, and Santa Barbara in the United States (Lastrico and Monge, E., 1972; Olsen, 1972). Vancouver and Victoria in Canada (Wuorinen, 1974) and Wellington, New Zealand (Adams, 1972), have also been mapped according to differences in seismic risk. So

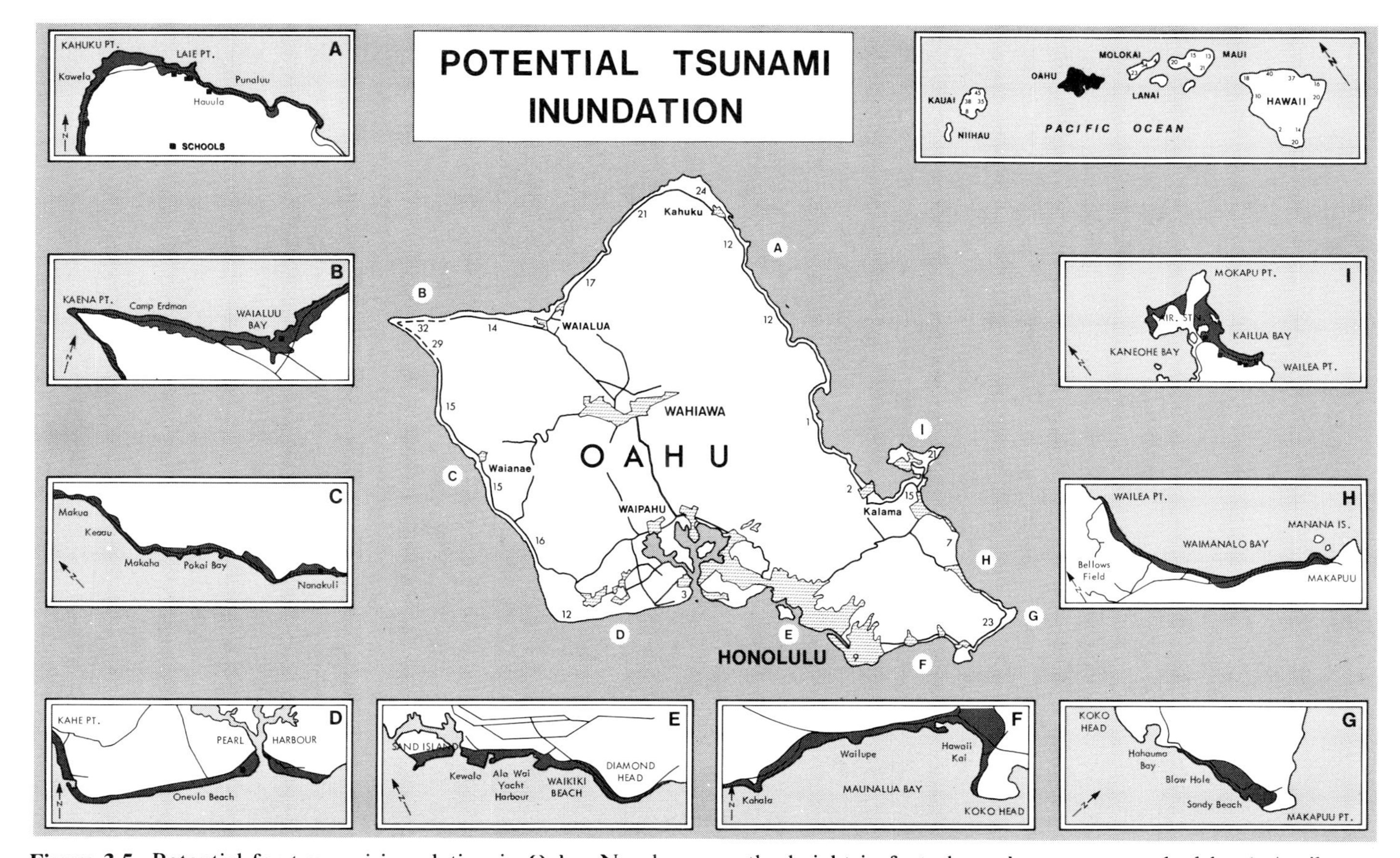

Figure 3.5. Potential for tsunami inundation in Oahu. Numbers are the height in feet above low water reached by 1 April 1946 tsunami (after Hawaiian Civil Defense and McDonald *et al.*, 1974).

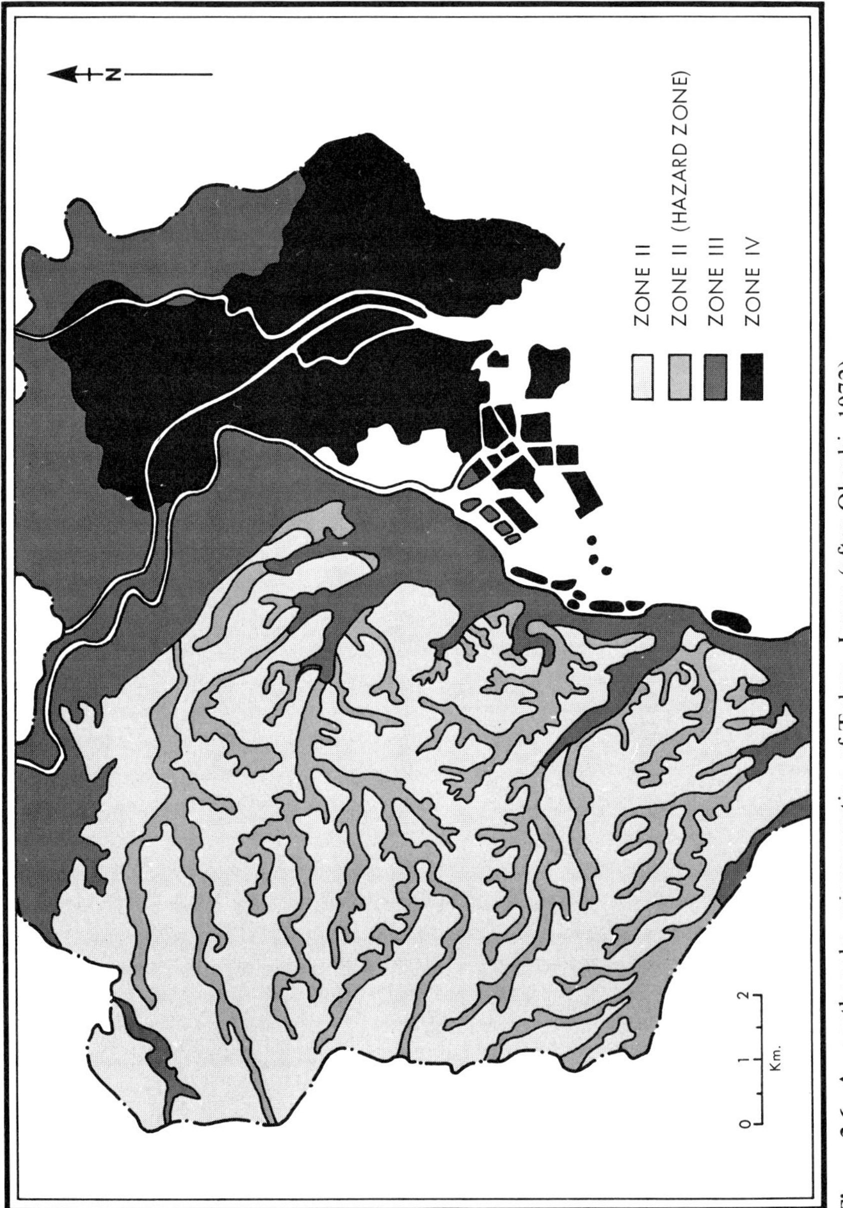

Figure 3.6. An earthquake microzonation of Tokyo, Japan (after Ohsaki, 1972).

has Tokyo, Japan (Ohsaki, 1972) as shown in Figure 3.6. Such maps permit spatial variations in the intensity of seismic ground disturbance to be predicted for earthquakes of differing magnitudes. Similarly, in April 1975, the Government of Canada announced a $20 million federal–provincial flood hazard mapping program for more than 200 urban and rural regions that are prone to flood damage. Its intent is to reduce development in known hazard areas. Maps produced to date show the probable extent of the 20-year and 100-year floods, and any other major inundation levels experienced recently in the area. Similar single hazard–multiple purpose maps include avalanche microzonations for parts of Colorado (Ives, Mears, Carrara, and Bovis, 1976) illustrated in Figure 3.7, and

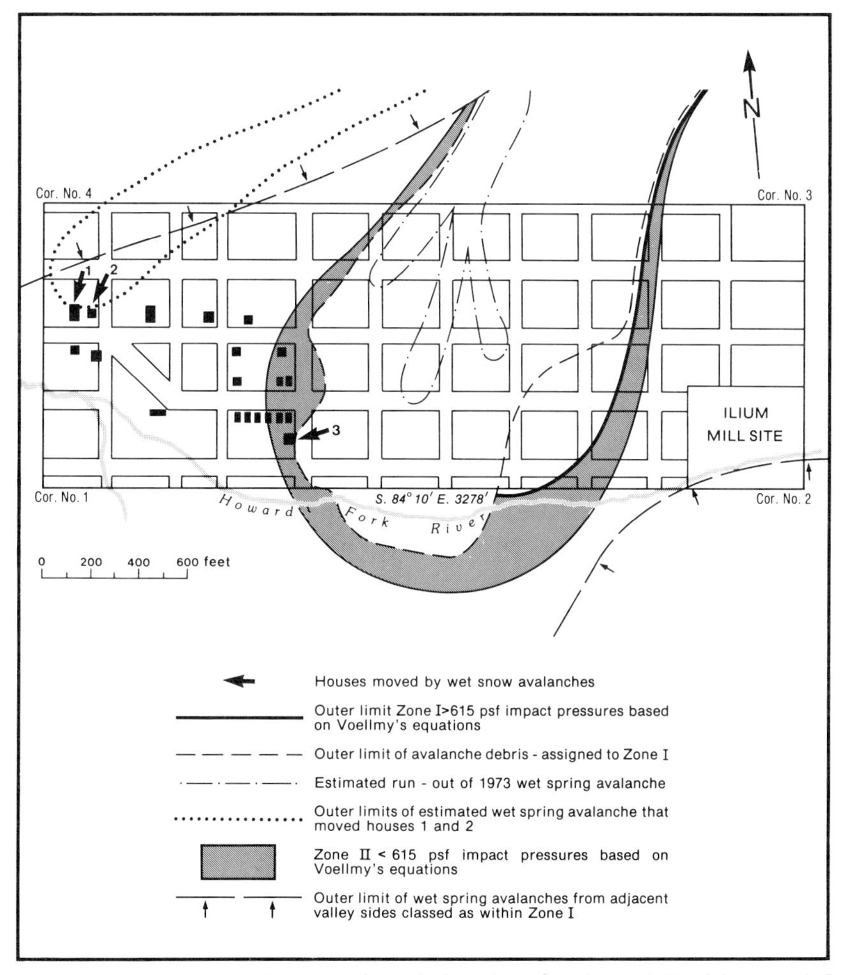

Figure 3.7. Avalanche microzonation, Colorado (after Ives, J.D., Meares, A.I., Carrara, P.E., and Bovis, M.J., 1976. Natural Hazards in Mountain Colorado, *Annals, Association of American Geographers* **66(1)**.

storm surge risk maps produced for the Galveston, Texas, area (Figure 3.8).

Multiple Hazard–Multiple Purpose Mapping

Certainly the most useful microzonations from a planning point of view are the multiple hazard–multiple purpose maps. These seek to show the total risk picture as a basis for rational locational decisions. To produce such maps, a common unit of comparison must be established for all disaster agents, such as the fatality or injury potential, probable stress caused, or the dollar losses to be expected. Since such mapping is likely to be computerized, the production of a series of maps, using all three of these units of measurement, is almost as simple a procedure as that involved in the use of a single basis of hazard comparison. Where a selection must be made, monetary units have certain advantages over other alternatives since they allow benefit–cost relationships to be established and permit potential losses due to risk to be compared with other locational costs, such as those connected with transportation. However, as has already been pointed out, there are many difficulties associated with assigning a monetary value to life loss and injury.

To produce multiple hazard–multiple purpose microzonations, individual disaster agents must be evaluated independently, using the techniques previously described, to produce a series of single hazard–multiple purpose microzonations. For each hazard mapped in this way the anticipated annual, or other time period, dollar losses per unit area are then calculated. The final total risk map is then the summation of all such values.

As early as 1971, Hewitt and Burton proposed an all-hazards-at-a-place research design and applied it to an area centered on London, Ontario. This technique set out to describe the whole spectrum of types of damaging events in the region and their aggregate significance in such terms as statistical measures and relationships. This study involved analysis of nine hazards: floods, hail, droughts, heavy rainfall, high wind, freezing rain, snowfall, tornadoes, and hurricanes. Each was portrayed in terms of its magnitude, return period, and past damage record. Nevertheless, no map combining the resultant total risk was ever produced.

A comparable study was conducted by the Puget Sound Governmental Conference for risks resulting from floods, earthquakes, windstorms, wildland fires, and volcanic activity. Maps delineating all hazards were to be prepared at a scale of 1:125,000, with levels of severity to be shown in two or more categories. Each hazard map was designed to serve as input data for a detailed risk analysis procedure, which produced a total loss potential figure. Unfortunately. budgetary limitations terminated the study

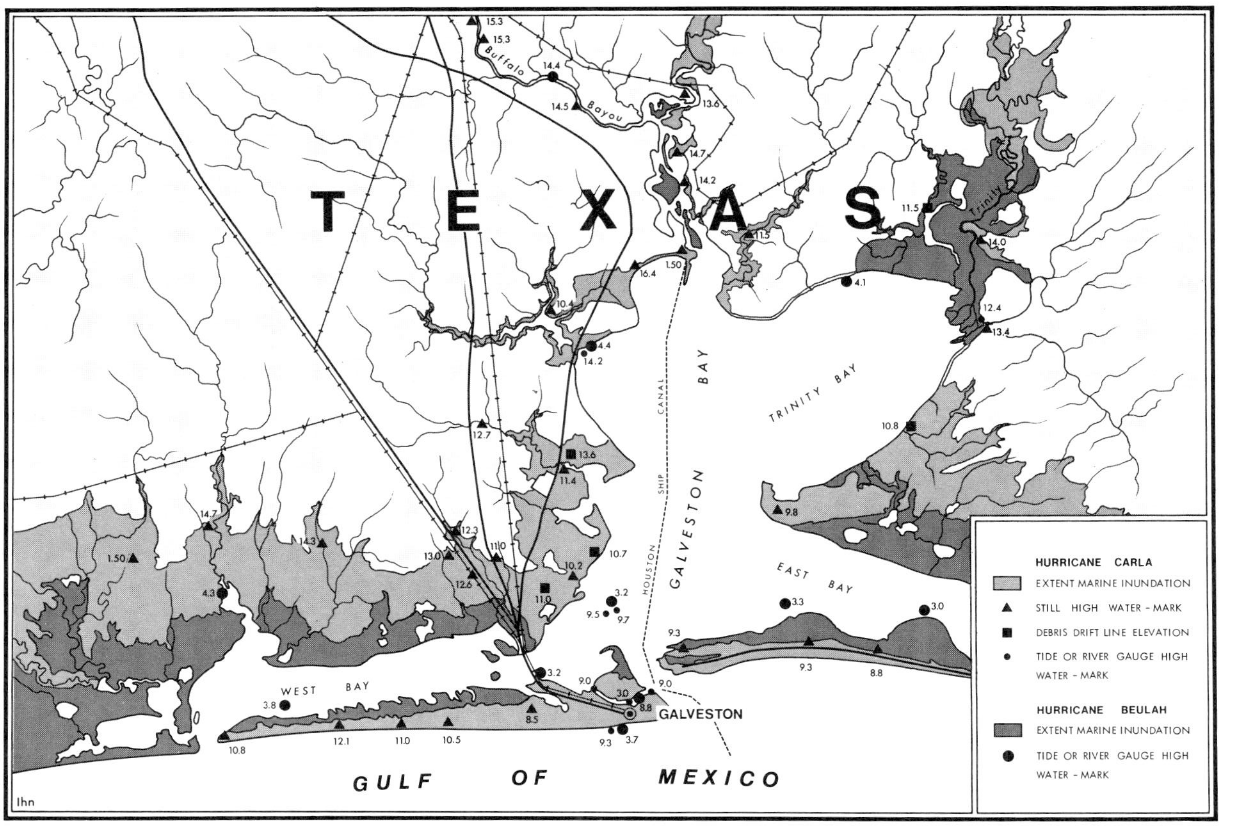

Figure 3.8. Storm surge microzonation of the Galveston, Texas area (after Bureau of Economic Geology, University of Texas at Austin).

at this point (Puget Sound Governmental Conference, 1975). Again, no overall total risk map was prepared.

One of the most comprehensive attempts to assess spatial variations in risk as an aid to policy making was the *Urban Geology Master Plan for California* (California Division of Mines and Geology, 1973). This three-year mapping project was completed in 1973. It was conceived as a tool for assisting local governments to respond to the risks posed by nine geological hazards and as an aid in the conservation of mineral deposits. The first phase of the study involved the preparation of statewide maps showing the location and degree of severity of ten geological problems. Such single hazard–one purpose microzonations were compiled at a scale of

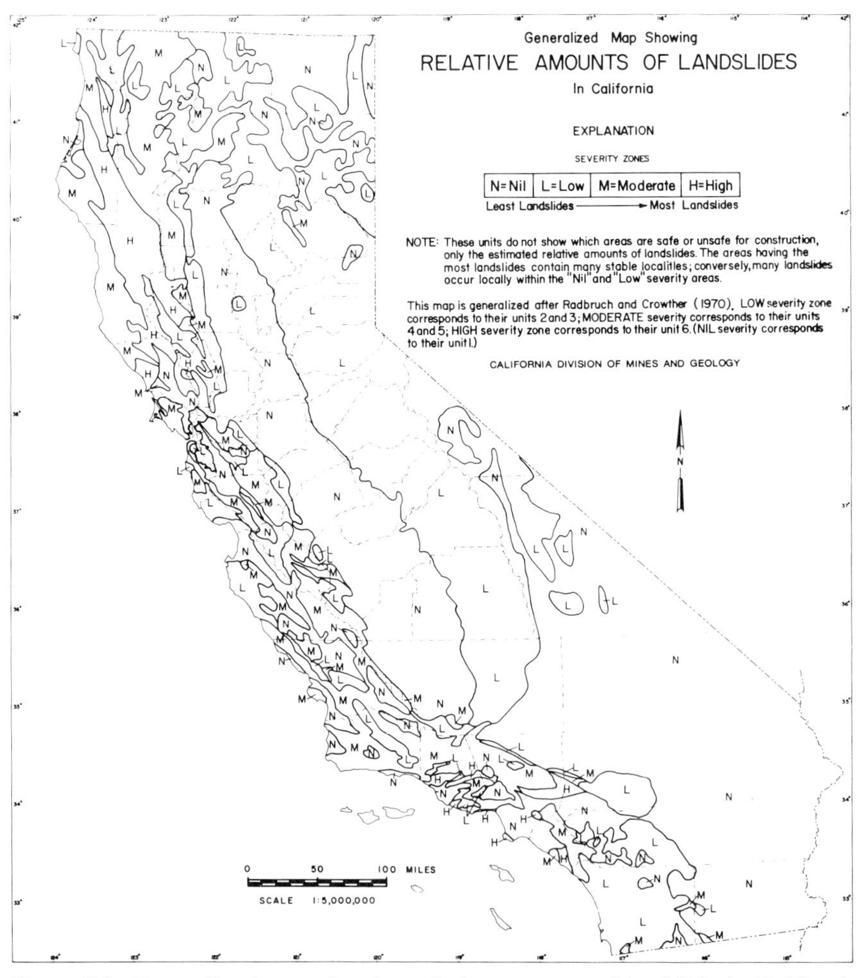

Figure 3.9. Generalized map showing relative amounts of landslide in California (California Division of Mines and Geology, 1973).

1:1,000,000 and published at one of 1:5,000,000. The hazards assessed in this way were earthquake shaking, fault displacement, landsliding, flooding, erosion, expansive soils, volcanic and tsunamagenic hazards, subsidence, and losses of mineral deposits beneath development. Each microzonation involved the use of four subdivisions, signifying the severity of the problem. The categories used were high, moderate, or low and none or not rated. Their application is shown in Figures 3.9 and 3.10.

Each of the microzonations was then combined into one composite map. This used a numerical coding system which allowed the severity levels for all of the ten problems to be shown in every 7.5 minute quadrangle in California. For example, a quadrangle might be depicted as

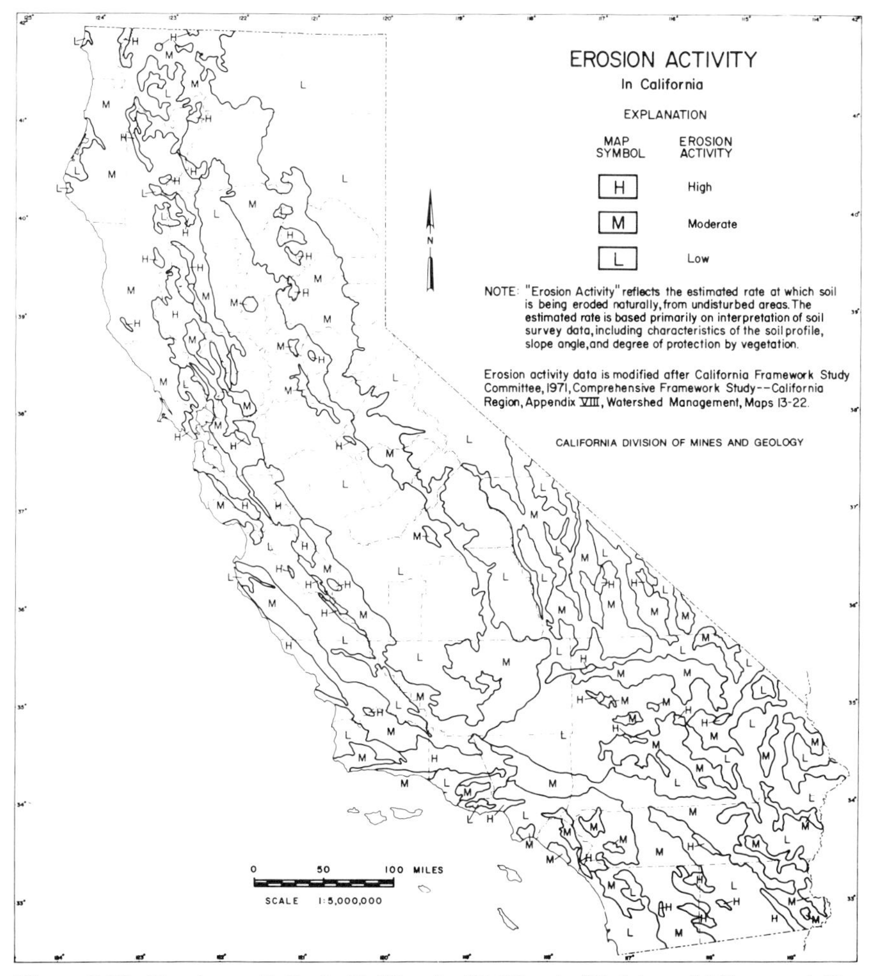

Figure 3.10. Erosion activity in California (California Division of Mines and Geology, 1973).

having high (3) earthquake shaking, flooding, and volcanic eruption potential; moderate (2) risk from fault displacement, landsliding, and subsidence, and low (1) loss probability from tsunamis and erosion. A blank in the appropriate position in the quadrangle would further illustrate that damage from expansive soils and the unnecessary loss of mineral resources was not an issue.

In an effort to calculate the economic significance of each geological hazard, a hypothetic urban area termed the "urban unit" was designed. This was envisioned as having population of 3000 and an infrastructural value of $90 million. In assessing total loss, each individual had a value of $75,000 placed on his life. This urban unit was then hypothetically subjected to all of the three severity levels for one of the geological problems, to allow an estimate to be made of the property damage and life loss that would result. Recurrence intervals for each severity level for all problems was also predicted. This information permitted annual per capita losses to be calculated. To illustrate, these amounted to $31 in high, $27 in moderate, and $14 in low severity earthquake shaking microzones (Table 3.3). For the purpose of the study these losses were termed geology points (GP).

In order to reflect the higher possible total losses and social significance of certain catastrophic events, a disaster factor (DF) was assigned to some hazards. Those such as landslides with the potential for taking between 1and 10 lives in a single event were given a DF value of 1.1. If 1 to 100 fatalities might result, as in an earthquake of moderate magnitude, a factor of 1.5 was used; if between 101 and 1000 deaths could occur the DF value assigned was 2. Higher severity events, such as a major earthquake, were given a factor of 3.

Table 3.3. California Geology Points[a]

Problem	Severity		
	High	Medium	Low
Earthquake shaking	31	27	14
Loss of mineral resources	22	—	—
Landsliding	53	35	1
Flooding	290	96	—
Erosion activity	3	2	1
Expansive soils	3	2	0
Fault displacement	5	0.50	0.06
Volcanic hazards	57	10	3
Tsunami hazards	144	14	1
Subsidence	0.34	0.02	0

Source: California Division of Mines and Geology (1973).
[a]Values correspond to anticipated average annual per capita loss in dollars.

The *Urban Geology Master Plan of California* was concerned with the actual probability that damage would occur from such disaster agents. For this reason population distribution and future development potential were also examined. Two independent aspects of population characteristics, present level of population in an area, and the timing of future growth were considered. The population level was expressed in person-years exposure (PY), defined as the average number of residents projected to be in the area during the three decades considered in the study: 1970–1980, 1980–1990, and 1990–2000. The timing of this population growth was expressed as an immediacy factor (IF) which was weighted to give higher priority to loss reduction actions in grid cells faced with earlier urbanization.

The four variables that were thought to govern the magnitude and significance of risk (geology points, the disaster factor, person-years exposure, and the immediacy factor) were then multiplied to develop priority points for each quadrangle. It was argued that the greater the resulting figure, the higher the priority in disaster mitigation that should be given to the quadrangle involved. This was because a high value implied numerous destructive disaster agents interacting with rapid urbanizations, a combination which would probably lead to high, yet often avoidable, losses. Table 3.4 illustrates a ranking of sample areas on such a priority point basis.

Three other independent approaches were used to help set disaster mitigation priorities. These included the ranking of hazards according to the potential damage for which they are likely to be responsible (Table 3.5.). In addition, hazards were ranked according to the ease and efficiency with which they could be reduced (Table 3.6). Attention was also paid to the possibility of emphasizing those disaster agents with the highest benefit:cost mitigation ratios (Table 3.7). These three approaches each resulted in distinct hazard rankings, illustrating that the setting of mitigation priorities in California is no simple matter. Nevertheless, such a step is clearly necessary since the study permitted estimates to be made of total losses for the state as a whole for the period 1970–2000. These were in excess of $55 billion. It was argued that if all feasible loss reduction methods were implemented, this figure could be reduced by over $38 billion (Table 3.8). In the case of every hazard, the benefit:cost ratio of such an approach to risk was positive.

This methodology has recently been refined and applied in far more detail to a much smaller area (Wuorinen, 1979). In the Saanich Peninsula, British Columbia, Wuorinen mapped the spatial distribution of threat from individual hazards, such as coastal and soil erosion, flooding, and earthquakes. In the case of earthquakes, for instance, three hazard zones were established and the frequency with which seismic events in each reached specified intensities was established by consulting the Canadian National Building Code.

Table 3.4. Priority Points for Disaster Mitigation Studies, Some Examples

Priority number	Quadrangle	County	Priority points
Southern California			
1	Laguna Beach NE 1/4	Orange	50,500
2	Orange SE 1/4	Orange	48,400
3	Orange NE 1/4	Orange	45,200
4	Newhall SW 1/4	Los Angeles	42,700
5	Tustin SW 1/4	Orange	37,400
6	Orange SW 1/4	Orange	36,700
1	Newhall SE 1/4	Los Angeles	36,500
8	Tustin NE 1/4	Orange	36,200
9	Laguna Beach NW 1/4	Orange	34,950
10	Yorba Linda NW 1/4	Los Angeles and Orange	34,700
Northern California			
1	Morgan Hill SE 1/4	Santa Clara	22,650
2	Niles SW 1/4	Alameda	21,300
3	Palo Alto NW 1/4	San Mateo	20,100
4	Palo Alto NE 1/4	San Mateo and Santa Clara	19,600
5	San Mateo SW 1/4	San Mateo	17,600
6	Morgan Hill SW 1/4	Santa Clara	17,000
7	Calaveras Reservoir SW 1/4	Santa Clara	16,300
8	Benicia NW 1/4	Contra Costa and Solano	16,100
9	San Jose East SW 1/4	Santa Clara	14,200
10	Santa Teresa Hills NE 1/4	Santa Clara	13,900

Source: California Division of Mines and Geology (1973).

To estimate associated losses it was also necessary to know the value of the land and improvements affected by such events. To calculate the average value of one hectare of developed land in the Greater Victoria area, financial data were obtained from the B.C. Assessment Authority. These had originally been collected for taxation purposes (Table 3.9). Independent values for land and improvements (public and private) were required for the production of total risk maps since different disaster agents result in losses to various combinations of investments. To illustrate, earthquakes generally affect improvements only, while coastal erosion destroys both the land and the improvements thereon.

It was also essential to know what percentage of this property would suffer damage in earthquakes or other disaster agent impacts of differing intensities. Steinbrugge (1970) has already established such damage ratios, based on the cost of repairs for seismic events. These were deter-

Table 3.5. Ranking of Hazards by Total Potential Damage

Geologic problem	Total loss, 1970–2000 ($ billions)
1. Earthquake shaking	21
2. Loss of mineral resources	17
3. Landsliding	10
4. Flooding	6.5
5. Erosion activity	0.5
6. Expansive soils	0.15
7. Fault displacement	0.08
8. Volcanic hazards	0.05
9. Tsunami hazards	0.004
10. Subsidence	0.003

Source: California Division of Mines and Geology (1973).

Table 3.6. Ranking of Hazards by Magnitude of Possible Loss Reduction

Geologic problem	Possible loss reduction 1970–2000 ($ billions)
1. Loss of mineral resources	15
2. Earthquake shaking	10
3. Landsliding	9
4. Flooding	3
5. Erosion activity	0.4
6. Expansive soils	0.15
7. Tsunami hazards	0.04
8. Subsidence	0.013
9. Fault displacement	0.013
10. Volcanic hazards	0.008

Source: California Division of Mines and Geology (1973).

Table 3.7. Ranking of Hazards on the Basis of the Benefit:Cost Ratio of Performing Loss Reduction Measures

Geologic problem	Benefit:cost ratio
1. Loss of mineral resources	170:1
2. Expansive soils	20:1
3. Landsliding	9:1
4. Earthquake shaking	5:1
5. Volcanic hazards	5:1
6. Fault displacement	1.7:1
7. Subsidence	1.5:1
8. Tsunami hazards	1.5:1
9. Erosion activity	1.5:1
10. Flooding	1.3:1

Source: California Division of Mines and Geology (1973).

Table 3.8. Projected Losses due to Geologic Problems in California, 1970–2000 (Estimated)

Geologic problem	Projected total losses, 1970–2000, without improvement of existing policies and practices	Possible total loss reduction 1970–2000, applying all feasible measures		Estimated total cost of applying all feasible measures, at current state of the art, 1970–2000		Benefit:cost ratio if all feasible measures were applied and all possible loss reductions were achieved, 1970–2000
		Percent of total loss	Dollar amount	Percent of total loss	Dollar amount	
Earthquake shaking	$21,035,000,000	50[a]	$10,517,500,000	10	$2,103,500,000	5
Loss of mineral resources	17,000,000,000	90	15,000,000,000	0.53	90,000,000	167
Landsliding	9,850,000,000	90	8,865,000,000	10.3	1,018,000,000	8.7
Flooding	6,532,000,000	52.5	3,432,000,000	41.4	2,703,000,000	1.3
Erosion activity	565,000,000	66	377,000,000	45.7	250,000,000	1.5
Expansive soils	150,000,000	99	148,500,000	5	7,500,000	20
Fault displacement	76,000,000	17	12,600,000	10	7,500,000	1.7
Volcanic hazards	49,380,000	16.5	8,135,000	3.5	1,655,000	4.9
Tsunami hazards	40,800,000	95	37,760,000	63	25,700,000	1.5
Subsidence	26,400,000	50	13,200,000	65.1	8,790,000	1.5
Total	$55,324,580,000	69	$38,411,695,000	11.2	$6,215,645,000	6.2

Source: California Division of Mines and Geology (1973).

[a]90 percent reduction of life loss.

Table 3.9. Value of Land and Improvements

	Land area (ha)	Land ($)	Private improvements ($)	Total improvements ($)	Grand total ($)
Victoria	1,878	152,033,160	280,137,800	348,146,273	500,179,433
Oak Bay	1046	261,464,089	182,234,550	227,793,198	489,257,287
Esquimalt	631	222,506,880	141,889,734	235,153,334	457,660,214
Total	3555	636,004,129	604,262,084	811,092,805	1,447,096,934
Value/ha	—	178,879	169,975	228,123	407,002

Source: B. C. Assessment Authority (1978); Wuorinen (1979).

mined from damage suffered in the San Francisco–Oakland metropolitan statistical area but were considered fairly applicable to Victoria.

These three types of information, the value of land and its developments, the expected frequency with which seismic intensities are to be expected, and the associated potential damage ratios, were then used to calculate the total anticipated loss per hectare per year in each of the earthquake risk zones (Table 3.10). In the lowest risk zone, annual earthquake related losses might be expected to be $266 per hectare, rising to $767 in moderate risk and $1,817 in high risk areas. In a similar manner, the annual losses to be anticipated in low, medium, and high risk flood, coastal, or soil erosion zones were also calculated (Table 3.11). Summing these values for each hectare resulted in a map showing total annual expected disaster losses for the Greater Victoria area (Figure 3.11). The data collected demonstrated that for the four natural hazards considered, average annual disaster losses per hectare ranged from $156 to some $4,600.

Table 3.10. Earthquake Losses by Zone

	Zone 1		Zone 2		Zone 3	
Annual frequency	MMI	Loss/ ha ($)	MMI	Loss/ ha ($)	MMI	Loss/ ha ($)
.150	III	—	IV	—	V	70
.065	IV	—	V	30	VI	141
.036	V	17	VI	78	VII	413
.015	VI	32	VII	172	VIII	564
.0081	VII	93	VIII	305	IX	447
.0033	VIII	124	IX	182	X	182
Total loss/ ha/year		266		767		1817

Source: Wuorinen (1979).

Table 3.11. Summary of Potential Annual Dollar Losses per Hectare for the Greater Victoria Area

Hazard	Zone 1	Zone 2	Zone 3	Zone 4
Earthquake	266	767	1817	—
Flooding	0	227	910	—
Surface erosion	0	3	9	—
Coastal erosion	0	960	1920	2880

Source: Wuorinen (1979).

Such multiple hazard–multiple purpose maps can, of course, be prepared for as many disaster agents as a region or municipality is subjected to. These hazards may be natural or man-made. Once their impact has been reduced to a common unit of measurement, risks from any source can be summed to give a comprehensive overview of potential losses. After such maps have been produced the next issue to be resolved is that of risk acceptability. The safety plan coordinator, his committee, and perhaps the general public must seek to agree on acceptable risk standards. These should reflect the careful balancing of potential losses and benefits.

Acceptable Risk

Risk to Life

Any microzonation will demonstrate spatial variations in risk. Some areas will be less hazardous than others. Nevertheless, regardless of location there is always risk. The issue then becomes one of deciding what level of risk is acceptable, to whom and for which activities. In the preceding chapter, the issue of the level of acceptable risk to life has been discussed in some detail. It has been shown that one approach to this problem is to use two limits to acceptable risk, of death from disease and from natural hazards, to set safety standards for the control of land use. The authors of the disaster mitigation plan for the Central Puget Sound Region (Puget Sound Council of Governments, 1975) argued that the higher the locational benefits accruing to a region from an activity, the greater should be the acceptable level of associated risk. Locational benefits were defined as "the total social benefits accruing to the region less the benefits which could be gained by locating the activity outside known hazard areas." The implied differences in benefits are caused by extra distance to markets, longer transportation routes, and sunk costs such as those of site preparation. When benefits are measured in this manner, those activities with the most restrictive siting requirements, such as port facilities and river gravel

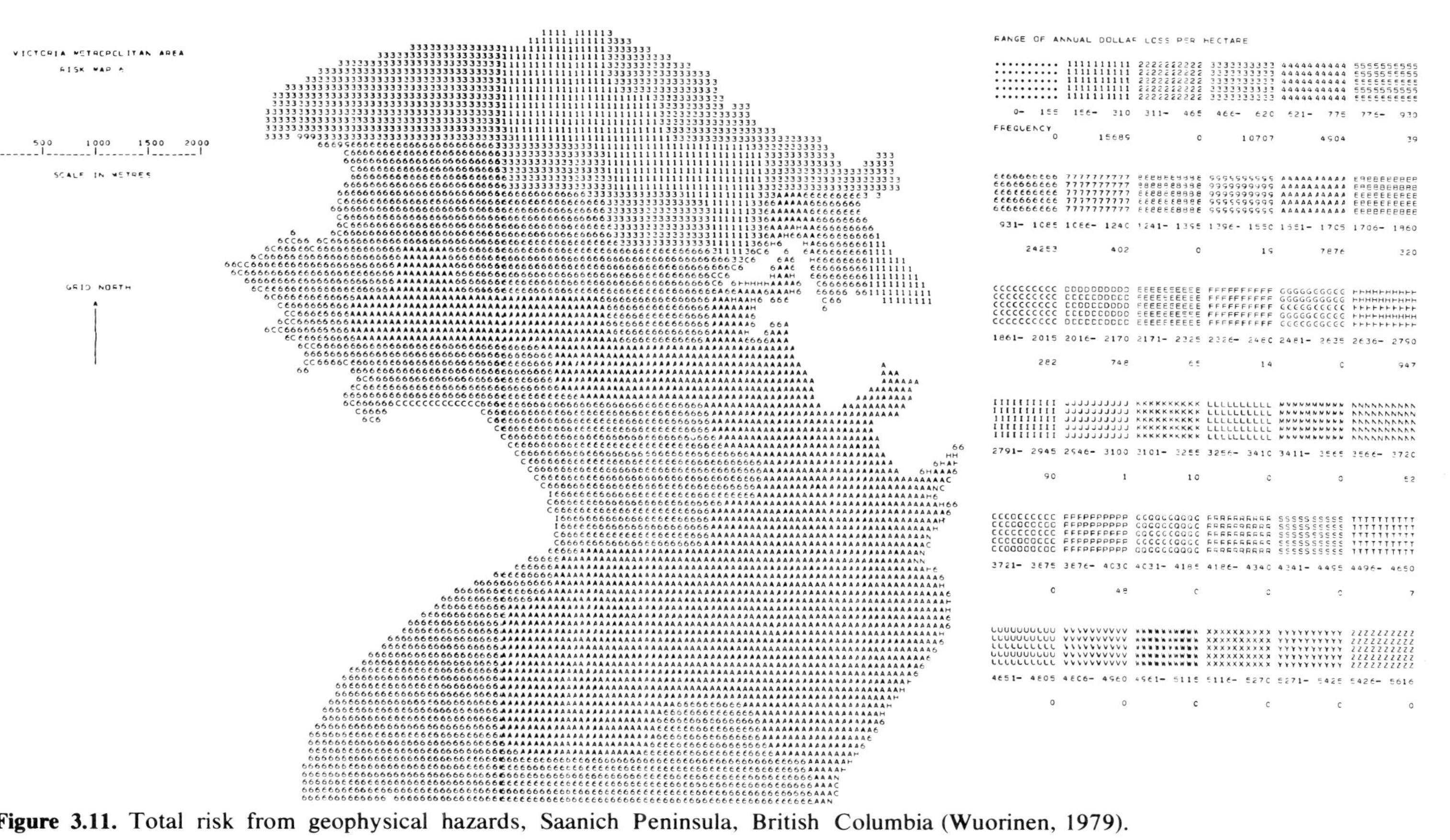

Figure 3.11. Total risk from geophysical hazards, Saanich Peninsula, British Columbia (Wuorinen, 1979).

extraction, also have the greatest locational benefits. This follows because these industries would be entirely precluded if they were not allowed to locate on a very narrow range of sites.

It was therefore argued that the more restrictive the locational demands of a particular land use, the greater the risk to life that society should tolerate. Using this concept, the standards of unacceptable personal risk shown in Table 3.12 were prepared. This table takes serious injuries as well as fatalities into consideration by using the generally accepted 10:1 conversion ratio, major injury being ten times as common as death. The average annual death and injury rate from disease and natural disaster was then used to set limits of unacceptable personal risk. Naturally, any safety committee is free to consider alternative safety standards.

Risk to Property

Growth within high risk areas should also be regulated on a benefit basis. Two questions must be squarely addressed to permit this. First, it must be established whether or not there is a favorable quantitative risk–benefit relationship including the cost of disaster-related activities. Normally, if benefits outweigh expected costs by an acceptable margin, development should be allowed. Second, the total amount of property at serious risk must be assessed to ensure that only a realistic proportion of total growth is taking place in high hazard areas. This requires the setting of

Table 3.12. Standards of Unacceptable Personal Risk

Maximum average annual percent casualties	Level of unacceptable risk	Land use activity	Typical uses
.0001	Greater than the average casualty rate due to natural disasters	Land use activities having little or no locational dependence	Most ordinary land use
1	Greater than the average casualty rate due to disease	Land use activities having strong locational dependence	Extractive industries such as sand and gravel mining; water related activities such as port activities

Source: Puget Sound Council of Governments (1975b).

an upper limit to total growth or maximum densities permitted in hazardous locations.

The level at which property risk becomes unacceptable generally depends upon the socioeconomic costs of disaster and on the size of the benefits accruing from the property in question (Puget Sound Council of Governments, 1975b). For these reasons extremely high risk standards are required for certain lifeline facilities, since these often impose great social and economic costs should they fail. Examples of this type of land use include communication centers, regional power intertie systems, chemical factories, and nuclear power plants. Any unnecessary risk to such uniquely vital or dangerous developments may result in enormous secondary damages should they fail. For this reason risk should be decreased to the greatest technologically feasible extent. The failure of necessary, but less dangerous, lifeline facilities such as fire and police services, hospitals, and the water supply system also carries lower, but still substantial, social costs. These services are vital in disasters and their facilities should be afforded sufficient integrity to guarantee their ability to function even during periods of emergency. Other lifeline facilities, courthouses, the city hall, and bus stations, for example, may have their services temporarily interrupted but should be restored on a priority basis. Slightly higher risk levels are generally tolerable for these facilities (Puget Sound Council of Governments, 1975b). Where a development has a substantial locational benefit and must, because of its nature, be built in a more hazardous zone, even higher risk levels can be allowed. In the case of farming, ski resorts, mines, quarries, and port facilities the offsetting benefits derived from the activity can, up to a point, be seen to validate the higher locational risks involved. These arguments suggest that safety plan coordinators and their committees might adopt the following allocation guidelines. They should attempt to prevent the development of vital or dangerous functions in hazard-prone areas as vigorously as possible and should seek to ensure the use of land on the basis of locational benefits.

An alternative method of approaching the problem of what is an acceptable level of risk would be for each local authority to produce a total risk map and to calculate the percentage of its area in zones of differing disaster potential. The highest risk 10% of the municipality, for instance, might then be set aside for open-air recreation and the lowest 5% for major lifeline facilities such as hospitals, disaster command posts, and power stations. Each alternative land use would then be assigned a particular portion of the risk spectrum within the municipality. This would mean that what was an acceptable level of risk for any type of development would vary with each jurisdiction. Politically, such an approach might be more realistic, since it is unlikely that any municipality that is predominantly within a high total risk zone would prohibit development for this reason.

Acceptable Total Risk

These guidelines do not address the problem of the maximum amount of high risk development any local authority can tolerate. This is a question of the timing of growth in zones of differing risk and of a region's ability to withstand disaster losses, which in turn reflects its wealth. Two approaches to this problem have been suggested (Puget Sound Council of Governments, 1975b). One is to determine the maximum face value of municipal bonds that local government would be willing to float to cover public losses, and the amount of insurance individuals would be prepared to pay for to offset their disaster losses. When this technique is applied, however, the figures obtained appear unrealistically low. An alternative, utilized in the Puget Sound area, was to estimate what the United States federal government was willing to pay out to disaster areas, without initiating loss reduction programs. This was calculated as roughly 0.12% of the total value of all reproducible, tangible assets susceptible to disaster damage in the United States. It was argued that this level of loss appeared to be very close to the limit of unacceptability, as evidenced by greatly increased federal disaster-related legislation, such as the National Flood Insurance Act and the Disaster Relief Act of 1974. It was believed that if local government permitted losses to mount much above this figure, by allowing increased development on high risk sites, they were inviting a loss of autonomy. In the Puget Sound region, damageable goods have an estimated value of roughly $40 billion, implying that the region's total annual average disaster loss potential should not be allowed to exceed $48 million. Although this standard may not be considered applicable in other municipalities or countries, it certainly provides a point of departure for discussions of the maximum amount of development permissible in any region's high risk microzones.

Table 3.13 summarizes the foregoing discussion of both unacceptable personal and property risks. Derived for use in the Puget Sound region, it also provides a series of standards that might be considered for adoption by any governmental body with the power to control land use. In addition, it provides a basis for the final presentation of total risk information. Ideally, risk microzonations should not seek to show average anticipated financial losses, or the stress likely to be experienced, but rather the suitability of each zone for a particular land use. This should be decided upon by comparing the estimated risk in any area with the risk standards set for each type of development. These in turn would reflect the stringency of the safety goals set by the community involved. Using hazard information and classes of unacceptable life and property risk like those illustrated in Table 3.13, such maps could be prepared for any region and would provide a very useful input into the early stages of the planning process.

Table 3.13. Standards of Unacceptable Property Risk Developed for the Puget Sound Area

	Level of unacceptable risk				
Maximum average annual percentage of casualties	Maximum annual total property damage	Maximum average annual percentage property damage	Description	Land use activity	Typical uses
.0001		Greater than lowest attainable level	Maximum structural integrity technologically attainable	Man-caused hazards whose failure might be catastrophic or highly unique lifeline facilities within the region which must continue to function during the emergency period	Class 1 lifeline facilities: emergency command centres, 911 communication centers, regional power intertie systems, man-caused hazards such as large dams or facilities using quantities of hazardous materials including nuclear power plants

Table 3.13. Standards of Unacceptable Property Risk Developed for the Puget Sound Area (*continued*)

Maximum average annual percentage of casualties	Level of unacceptable risk		Description	Land use activity	Typical uses
	Maximum annual total property damage	Maximum average annual percentage property damage			
.0001		Greater than .01	Sufficient structural integrity to ensure continued functioning and the lowest tolerable life risk value	Less unique but vital lifeline services of critical importance to the community's life saving ability which must be able to function during an emergency	Class 2 lifeline facilities: hospitals, blood banks, police and fire departments, water supply mains, critical freeway links
.0001	$48 million	Greater than .03	Sufficient structural integrity or redundancy to ensure rapid normalization of service and provide the lowest tolerable life risk value	Lifeline functions which may be interrupted but which must be restored to service immediately on a priority basis; man-caused hazards whose failure would be costly but less than catastrophic	Class 3 and 4 lifeline facilities such as communication and power networks, highway networks, nursing homes, jails, asylums, county court houses, and city halls; man-caused hazards such as

				natural gas, oil, and sewer lines
.0001	Greater than .05	Sufficient structural integrity to provide the lowest tolerable life risk value	Lifeline functions which may be interrupted but which must be restored during the economic recovery period on a priority basis	Major employers, transportation networks including airports, railways, and noncritical highway links
.0001	Greater than .30	Sufficient structural integrity to provide the lowest tolerable life risk value	Ordinary development having little or no locational benefit and neither an especially hazardous or vital use	Ordinary residential, manufacturing, commercial, or other ordinary uses
1	Greater than the annual location benefit	Sufficient structural integrity to provide a life risk no greater than the highest tolerable value	Development having strong location requirements and providing substantial benefit	Extractive industry such as sand and gravel mining and related industry such as port facilities

Source: Puget Sound Council of Governments (1975b).

Enforcing Unacceptable Risk Standards

If total risk mapping is to reduce losses and assist in the achievement of safety goals, then its use as a tool for guiding development must be supported by all local planning and government agencies. These bodies have a wide variety of techniques at their disposal with which they can influence the location of various types of land use. One of the methods used most frequently by government to regulate private development is zoning. The authority to guide land use in this manner is usually delegated to regional and local governments or to the administrations of towns and cities. Total risk maps and associated disaster simulations provide such institutions with a very useful base from which to develop zoning regulations. In these plans, high value development can be restricted to low risk locations. Where potential losses are presented in dollar values, any zoning exceptions should be based on benefit:cost ratios which clearly demonstrate that the greater expected damages are justified by the larger financial returns from a high risk development siting. Such a total risk map is being prepared for controlling future growth in the Philippines for the Greater Manila region (J.P. Lévy, 1976, personal communcation,). In North America, risk is often taken into account in the preparation of zoning ordinances but consideration is normally restricted to one or perhaps two hazards. The Northeastern Illinois Planning Commission (1964), for example, has prepared a model flood plain zoning ordinance for the assistance of county and local governments. Several of these counties including Cook, Du Page, and Lake have adopted it or similar versions (Sheaffer, Ellis, and Spieker, 1969). This ordinance restricts residential development in high risk locations; provides for permanent floodway channels through the acquisition of rights-of-way, including easements for maintenance and improvements; and requires the floodproofing of buildings within high hazard areas. Much of the pressure for the adoption of such regulations has come from the Metroplitan Sanitary District that has refused to issue sewer permits for new connections to its system, unless these criteria had been met. Similarly, the town of Portola Valley, in close proximity to the San Andreas Fault in California, is using microzoning maps to enforce land use restrictions including building setbacks to reduce the earthquake hazard (Mader, Danehy, Cummings, and Dickinson, 1972). Single hazard zoning is also in use elsewhere. Maps of Javanese volcanoes, for example, clearly delineate a prohibited zone in which no construction is allowed and into which the population is prevented from entering during an eruption (Figure 3.4).

Building permits can also be used with great effect to influence development. Microzoning maps, delineating variations in landsliding potential, almost invariably show that the highest risks occur in areas of steep gradients and great precipitation, particularly if building has taken place on weathered and well-jointed or fractured bedrock in seismically unstable areas. Los Angeles, because of its relatively weak rocks and steep slopes, has had a long history of landslide damage, most of which occurs during periods of intense rainfall. Before 1952 no studies of landslide potential were required before building was permitted; after this date until 1962 a moderately effective grading ordinance was in effect, while from 1963 to the present time detailed geomorphological site surveys have been required before any construction was allowed. During the wet year 1969, of the 10,000 hillside lots developed prior to 1952, 1040 failed, causing $3,300,000 million damage. Of the 27,000 sites built on during the period 1952–1962, 350 were also damaged, causing losses of $2,767,000. In contrast, of the 11,000 home sites developed since 1963, only 17 showed any damage. These figures indicate that the careful application of landsliding microzonations and site stability studies in Los Angeles have reduced the percentage of lots being adversely affected by landsliding from 10.4 to 0.15% (Jahns, 1969).

Financial institutions, both private and public, can exert a powerful influence over the siting of developments. Banks and credit unions would rather not invest their capital in buildings in high risk areas, and such structures should also carry higher insurance rates. In the Chicago metropolitan area, the Veterans Administration and the Federal Housing Administration routinely use the United States Geological Survey's flood-hazard maps to determine whether or not to finance developments. High risk projects will not be funded. A similar policy is followed by many banks and saving and loan companies. In the same way private utility companies can influence urban growth by extending or refusing to extend their gas and electricity lines into high risk zones where they can anticipate repeated damage to these facilities.

There are many reasons for the public purchase of land, for example, park acquisitions, the development of green belts, or provision of municipal car parking facilities outdoors. Most of these activities represent low investment and are a suitable land use for high risk zones. For several years the Du Page County Forest Preserve District has been engaged in a long-range program of land acquisition, designed to develop a major green belt along the West Branch of the Du Page River. Guidelines for purchase have been provided by the United States Geological Survey's flood-hazard maps (Sheaffer, Ellis, and Spieker, 1969). At Cowichan Head, Vancouver Island, rapid coastal erosion is occurring, threatening cliff-top residents who are attempting to retard it by building a seawall and by dumping rubble onto the beach. Should their efforts prove successful, Island View Beach, a major recreational area to the north, would lose its source of sand supply and rapidly deteriorate. The author has suggested that areas of the eroding cliff be purchased by public funds and encouraged to retreat, so nourishing the recreational beach fed by longshore drifting (Foster and Norie, 1978).

Ill-advised siting of public facilities often encourages the development of high risk zones. Highways commonly use flood plains and avalanche-prone valleys. The proper planning of access roads and junctions can tend to discourage the rapid development of high risk sites. Sewage treatment plants and sanitary landfills should also be located on low risk sites, since they can be responsible for secondary disasters if damaged by certain types of disaster agent.

Opposition to Microzonation

One major obstacle faced by those wishing to use risk maps as a basis for planning has been their high cost of preparation. Naturally this varies with the size of the region being microzoned, the scale of the resulting map, the diversity of hazards being considered, the prior availability of information, and the techniques employed. Costs also vary with the time allocated to the project and the expertise of those involved in its completion. Some data are available. An area of 200 square miles was mapped in Utah at a scale of 1:24,000 to determine its susceptibility to landsliding and fault slippage. Field survey costs and mapping totaled $24,000 or $120 per square mile. Flood hazard mapping in Metropolitan Chicago costs between $106 to $176 per square mile to complete, while California landslide maps at a scale of 1:6000 have required as much as $1000 per square mile to produce (Baker and McPhee, 1975). Wuorinen (1979) argues that adoption of his computerized total risk mapping methodology could reduce these figures.

Once the spatial distribution of risk has been widely recognized and standards of unacceptability adopted, changes in land values will occur. This process naturally leads to opposition from the owners of adversely affected high risk sites. The former planning director of Anchorage, Alaska, has described the enormous pressures that can arise to permit building on high risk locations (Schoop, 1969). Immediately after the 1964 earthquake, microzonations were prepared to ensure that Anchorage was rebuilt in a logical fashion and that high risk landslide prone areas remained undeveloped. In Schoop's own words:

> Then the money started breaking loose—federal funds, insurance funds, Small Business Administration loans, new capital. Up until then the task of rebuilding had looked grim. It had looked like it was a matter of carefully husbanding scarce resources, as Anchorage had been used to doing. It looked as though it would be a slow and painful process, one which should be done right so it wouldn't have to be done agian. The planners' and the geologists' proposals of how to accomplish this were being listened to.
>
> But somewhere, and I reckon it about the 45th day after Good Friday, a louder noise was heard. It was the scramble of disaster-affected owners—which was almost everyone, it seemed—to get their projects

PLATE 6. Turnagain Heights, Alaska, a landslide triggered by the March 1964 earthquake caused great damage to this residential area. The area has not been resettled (U.S. Army photograph).

going, to get in *first*. There was a scramble to get moving, lest there not be enough money, lest there not be enough labor and materials, or lest the market for this or that economic use become saturated too fast—to the point that even though one had the money, he couldn't cash in on it.

It became a really heady pell-mell, and the last ones to be listened to as that crescendo mounted now were the planners and the geologists. The professional disaster followers were, indeed, right about the rate of reconstruction.

Whereas in the wake of the disaster in the first days of the agony the request was, "Planner, let's do it right," now the refrain was, "Planner *get out of the way*."

The task force maps that federal forces, as well as local ones, had prepared indicating hazard areas and what ought to be avoided were ignored. A future governor of Alaska was to show how you go in and rebuild by building a very, very flashy new hotel—a high rise hotel—right in one of those hazard areas. Areas which should have been left open, or at least held down to single-family density, according to the task force maps, were being rebuilt with apartment houses. The political pressure mounted on even the federal task force—which was not immune—to water down the warnings in these task force ratings of the haz-

> ard areas. Even though they were being ignored, they still were an obstacle to getting in there. And they were, in fact, watered down—*twice*. The names were changed and meanings were changed so that private capital was less afraid to come in, even though as I have demonstrated already, it was not afraid. More could now come in, and it could even be opened up for federal funds to be spent one way or another in these areas.
>
> Two years after the earthquake—two years—the remnants of those who cared about the future of Anchorage met at my house; we concluded that this tide was so strong it could not be reversed. It was going to have to run its course. That's when I finally did "get out of the way."

Schoop considered that this wild scramble to repeat the errors of the past is normal, a conclusion that he buttressed by pointing to the irrational reconstruction which followed the 1906 San Francisco earthquake.

Regardless of the strength of development pressures in high risk areas, economic studies almost invariably support the use of microzonation in directing construction. Detailed surveys of damage suffered by residents of California, for example, indicate that in high risk earthquake zones, annual per capita losses from this hazard are $31; losses being $27 in intermediate zones, and $14 in low risk areas. The cost of flooding is considerably greater, reaching a maximum in very prone areas of $290 per capita per year. Losses from landsliding are $53, $35, and $1 in zones of high, moderate, and low risk, respectively. Unless active disaster mitigation techniques are applied in California, losses of over $55 billion from these and other hazards can be anticipated by the year 2000 (Table 3.3). Clearly, planning based upon widespread hazard microzonation is essential in California (California Division of Mines and Geology, 1973). Many developers and government agencies there, as elsewhere, resist such a rational approach to disaster mitigation on the grounds that it is too costly. Yet this study indicates that when applied, the benefits of such measures (the saving of lives and prevention of property damage) often greatly outweigh the costs involved. Examples cited include a benefit:cost ratio of 20:1 for expansive soils, 9:1 for landsliding, 5:1 for earthquake shaking and volcanic hazards, 1.5:1, subsidence, tsunamis, and erosion, and 1.3:1 for flooding. These figures demonstrate that in the long run it is cheaper to plan to avoid disaster than to suffer it.

References

Adams, R. D. 1972. Microzoning for earthquake effects in the Wellington city area. *Bulletin of the New Zealand Society for Earthquake Engineering*, **5**:106–107.

Baker, E. J., and J. G. McPhee. 1975. *Land Use Management and Regulation in*

Hazardous Areas: A Research Assessment. Institute of Behavioral Science, University of Colorado, Boulder.

Beasley, R. P. 1972. *Erosion and Sediment Pollution Control.* Iowa State University Press, Ames, Iowa, 320 pp.

Bennett, J. G. 1963. Geo-physics and human history: New light on Plato's Atlantis and the Exodus. *Systematics,* **1:**127, 156.

Binnie, G. M., and M. Mansell-Moullin. 1966. The estimated probable maximum storm and flood on the Jhelum River—A Tributary of the Indus. *The Institute of Civil Engineers Proceedings of the Symposium on River Flood Hydrology,* **189:**210.

Brumbaugh, R. S. 1970. Plato's Atlantis. *Yale Alumni Magazine,* **33**(5):24–28.

California Division of Mines and Geology. 1973. *Urban Geology Master Plan for California.* Bulletin 198, Sacramento, California, 112 pp.

Canadian Press. 1975. Mercury pollution worry among North Indians. *Victoria Times,* October 16, p. 21.

Chorley, R. J., and P. Haggett. 1967. *Models in Geography.* Methuen, London, 816 pp.

Collier, E. P., and G. A. Nix. 1967. *Flood Frequency Analysis for the New Brunswick-Gaspé Region.* Technical Bulletin No. 9, Inland Waters Branch, Department of Energy, Mines and Resources, Ottawa.

Costa, J. E. 1978. Holocene stratigraphy in flood frequency analysis. *Water Resources Research,* **14(4):**626–632.

Council for Science and Society. 1977. *The Acceptability of Risks.* Barry Rose, London, 104 pp.

Crandell, D. R., and D. R. Mullineaux. 1967. Volcanic hazards at Mount Rainier, Washington. *United States Geological Survey Bulletin 1238.*

Crandell, D. R., and H. H. Waldron. 1969. Volcanic hazards in the Cascade Range. *In* R. A. Olson and M. M. Wallace (Eds.), *Geologic Hazards and Public Problems.* Region Seven, Office of Emergency Preparedness, Santa Rosa, California, pp. 5-18.

Fenge, T. 1976. Geomorphic aspects of sanitary landfill site selection. *In*: H. D. Foster (Ed.), *Victoria: Physical Environment and Development, 12, Western Geographical Series,* pp. 241–288.

Fosberg, M. A. 1973. New technology for determining atmospheric influences on smoke concentrations. *In: Proceedings of International Symposium on Air Quality and Smoke from Urban and Forest Fires.* Proceedings, National Academic Sciences, Washington, D.C., pp.148–159.

Foster, H. D. 1975. Disaster mitigation: A geomorphological contribution. *Emergency Planning Digest,* **2(5):**2–9.

Foster, H. D., and I. Norie. 1978. *Coastal Erosion and Resource Management: Two Case Studies from Vancouver Island, British Columbia.* Paper presented at the 1978 Annual General Meeting of the Canadian Association of Geographers, London, Ontario, May 23–27.

Fraser, C. 1966. *The Avalanche Enigma.* John Murray, London, 301 pp.

Galanopoulos, A. G. 1960. On the origin of the deluge of Deukalion and the myth of Atlantis. *Athenais Archaiologike Hetaireia,* **3**, 226–231.

Galanopoulos, A. G. 1964. Die ägyptischen Plagen und der Auszug Israels aus geologischer Sicht. *Das Altertum,* **10**, 131–137

Galanopoulos, A. G. 1969. Der Phaëthon—Mythus im Licht der Wissenschaft. *Das Altertum*, **14**, 158-161.

Geological Survey of Indonesia. Microzonation maps at a scale of 1:50,000 are prepared by the Volcanology Division, Djayadi Hadikusamo, Chief, of the Geological Survey of Indonesia.

Gumbel, E. J. 1958. Statistical theory of floods and droughts. *Journal of the Institute of Water Engineers*, **12**, 157–184.

Harriss, J. 1979. Inferno at Los Alfaques. *Readers Digest*, **114(686)**:54–58.

Hansen, W. R., and E. B. Eckel. 1966. The Alaska earthquake, March 27, 1964: Field investigation and reconstruction effort. *U.S. Geological Survey Professional Paper 541*.

Hewitt, K. 1970. Probabilistic approaches to discrete natural events: A review and theoretical discussion. *Economic Geographer Supplement*, **46**:332–349.

Hewitt, K., and I. Burton. 1971. *The Hazardousness of a Place: A Regional Ecology of Damaging Events*. University of Toronto Press, University of Toronto Department of Geography Research Series Publications, Toronto, Ontario, 154 pp.

Hodgson, J. H., 1964. *Earthquakes and Earth Structure*. Prentice-Hall, Englewood Cliffs, New Jersey, 166 pp.

Ives, J. D., A. I. Mears, P. E. Carrara, and M. J. Bovis. 1976. Natural hazards in mountain Colorado. *Annals of the Association of American Geographers*, **66(1)**:129–144.

Ives, R. L. 1962. Dating the 1746 eruption of Tres Virgenes Volcano, Baha California del Sur, Mexico. *Bulletin of Geological Society of America*, **73**:647-648.

Jahns, R. H. 1969. Seventeen years of response by the City of Los Angeles to geologic hazards. *In:* R. A. Olson and M. M. Wallace (Eds.), *Geologic Hazards and Public Problems*. Office of Emergency Preparedness, Region Seven, Santa Rosa, California, pp.283–295.

Kárník, V. 1972. Microzoning program within the UNDP-UNESCO survey of the seismicity of the Balkan region. *Proceedings of the International Conference on Microzonation for Safe Construction Research and Application. NSF, UNESCO*, University of Washington, ASE and Academy of Mechanics. Seattle, Washington, pp. 213–216.

Kates, R. S. 1970. *Natural Hazards in Human Ecological Perspective: Hypotheses and Models*. Natural Hazards Research Working Paper No. 14. University of Toronto Press, Toronto.

Lang, T. E., K. L. Dawson, and M. Martinelli, Jr. 1979. *Numerical Simulation of Snow Avalanche Flow*. Research Paper RM-205. Rocky Mountain Forest and Range Experimental Station, Forest Service, U.S. Department of Agriculture.

Lastrico, R. M., and J. Monge, E. 1972. Chilean experience in seismic microzonation. *Proceedings of the International Conference on Microzonation for Safer Construction Research and Application.* NSF, UNESCO, University of Washington, ASE and Academy of Mechanics, Seattle, Washington, pp. 231–248.

Lyon, J. 1979. Quoted in the *Victoria Times*, April 20. Those pretty red clouds were bearing leukemia, p. 55.

Macdonald, G. A. 1972. *Volcanoes*. Prentice-Hall, Englewood Cliffs, New Jersey.

Macdonald, G. A., F. P. Shepard, and D. C. Cox. 1947. The tsunami of April 1, 1946 in the Hawaiian Islands. *Pacific Science,* pp. 21–37.

Mader, G. G., E. A. Danehy, J. C. Cummings, and W. R. Dickinson. 1972. Land use restrictions along the San Andreas Fault in Portola Valley, California. *Proceedings of the International Conference on Microzonation for Safer Construction Research and Application.* NSF, UNESCO, University of Washington, ASE and Academy of Mechanics, Seattle, Washington, pp. 845–858.

Madole, R. F. 1974. Quaternary research methodologies applied to delimiting natural hazards in mountainous Colorado. *In: Quaternary Environments Proceedings of a Symposium.* York University–Atkinson College Geographical Monograph No. 5, pp. 79–98.

Magoon, O. T. 1966. Structural damage by tsunamis. *Proceedings of the American Society of Civil Engineers. Coastal Engineering Specialty Conference, 1965.* American Society of Civil Engineers, New York, pp. 35–67.

Marinatos, S. 1939. The volcanic destruction of Minoan Crete. *Antiquity,* **13**:425-439.

Northeastern Illinois Planning Commission. 1964. *Suggested Flood Damage Prevention Ordinance with Commentary.* Northeastern Illinois Planning Commission, Chicago, 28 pp.

Ohsaki, Y. 1972. Japanese microzonation methods. *Proceedings of the International Conference on Microzonation for Safer Construction Research and Application.* NSF, UNESCO, University of Washington, ASE and Academy of Mechanics, Seattle, Washington, pp. 162–182.

Olsen, P. G. 1972. Seismic microzonation in the City of Santa Barbara, California. *Proceedings of the International Conference on Microzonation for Safer Construction Research and Application.* NSF, UNESCO, University of Washington, ASE and Academy of Mechanics, Seattle, Washington, pp. 395–408.

Penning-Rowsell, E. C., and J. B. Chatterton. 1977. *The Benefits of Flood Alleviation: A Manual of Assessment Techniques.* Saxon House, Farnborough, England, 297 pp.

Perla, R. I., and M. Martinelli, Jr. 1976. *Avalanche Handbook.* Agriculture Handbook 489, U.S. Department of Agriculture, Forest Service.

Pitty, A. F. 1971. *Introduction to Geomorphology.* Methuen, London, 526 pp.

Puget Sound Council of Governments. 1975a. *Regional Disaster Mitigation Plan for the Central Puget Sound Region.* Puget Sound Council of Governments, 106 pp.

Puget Sound Council of Governments. 1975b. *Regional Disaster Mitigation Technical Study for the Central Puget Sound Region.* Puget Sound Council of Governments, 208 pp.

Renault, M. 1962. *The Bull from the Sea.* Pantheon, New York.

Ritter, J. R., and W. R. Dupre. 1972. Map published as Miscellaneous Field Study, U.S. Geological Survey, Reston, Virginia.

Salm, B. 1965. Contribution to avalanche dynamics. *Proceedings, International Symposium on Scientific Aspects of Snow and Ice Avalanches.* International Association of Scientific Hydrology Publication 69, pp. 199–214.

Schaerer, P. A. 1972. Terrain and vegetation of snow avalanche sites at Rogers Pass, British Columbia. *In:* O. Slaymaker and H. J. McPherson (Eds.), *Mountain Geomorphology.* Tantalus, Vancouver, B.C., pp. 215–222.

Schoop, E. J. 1969. Development pressures after the earthquake. *In:* R. A. Olson and M. M. Wallace (Eds.), *Geologic Hazards and Public Problems.* Region Seven, Office of Emergency Preparedness, Santa Rosa, California, pp. 229–232.

Sheaffer, J. R., D. W. Ellis, and A. M. Spieker. 1969. *Flood-Hazard Mapping in Metropolitan Chicago.* U.S. Geological Survey Circular 601-C, 14 pp.

Smithsonian Institution. 1971. *Natural Disaster Research Centers and Warning Systems: A Preliminary Survey.* Smithsonian Institution, Cambridge, Massachusetts.

Solov'yev, S. E. 1968. The Sanakh-Kad'yakskoye tsunami of 1788. *In: Problema Tsunami.* Akad. Nauk. SSSR Otdeleniye Nauk o Zemli, Sovet po Seysmologii, Moscow, pp. 232–237.

Steinbrugge, K. V. 1970. Earthquake damage and structural performance in the United States. *In:* R. L. Wiegel (Ed.), *Earthquake Engineering.* Prentice-Hall, Englewood Cliffs, New Jersey, pp. 167–226.

Steinbrugge, K. V., and V. R. Bush. 1965. Review of earthquake damage in the Western United States, 1931–1964. *In:* D. S. Calder (Ed.), *Earthquake Investigations in the Western United States.* U.S. Department of Commerce, Washington, D.C., pp. 223–256.

Sumi, K. and Y. Tsuchiya. 1976. *Assessment of Relative Toxicity of Materials—Toxicity Index.* National Research Council of Canada, NRCC 15367, DBR Paper 685.

Texas Coastal and Marine Council. 1978. *Model Minimum Hurricane-Resistant Building Standards for the Texas Gulf Coast.* The Texas Coastal and Marine Council, Austin, Texas.

Tudor, W. J. 1964. *Tsunami Damage at Kodiak, Alaska and Crescent City, California fom Alaskan Earthquake of 27 March 1964.* U.S. Naval Civil Engineering Laboratory, Port Hueneme, California, Technical Note N-622.

United Press International. 1979. 'Safety Tests' scattered radioactivity in Utah. *The Victoria Times,* March 2, p. 17.

U.S. Department of Agriculture. 1973. *Evapotranspiration and Water Research as Related to Riparian and Phraetophyte Management: An Abstract Bibliography.* Miscellaneous Publication 1324, 192 pp.

U.S. National Water Commission. 1973. *Water Policies for the Future.* Water Information Center, Port Washington, New York, 579 pp.

U.S. Office of Emergency Preparedness. 1972. *Disaster Preparedness.* Report to Congress, Executive Office of the President, 184 pp.

Vitaliano, D. B. 1973. *Legends of the Earth: Their Geologic Origin.* Indiana University Press, Bloomington, Indiana, 305 pp.

Water Survey of Canada. 1974. *Historical Streamflow Summary British Columbia to 1973.* Inland Waters Directorate, Environment Canada.

Welsh, M. S. 1973. Remote sensing and the assessment of coccidioidal hazards in arid regions. *Proceedings of the Fourth Annual Conference on Remote Sensing in Arid Lands.* University of Arizona, Tuscon, p. 125.

White, G. F. 1964. *Choice of Adjustment to Floods.* Research Series No. 93, Department of Geography, University of Chicago, Chicago.

Whitman, R. V. 1973. *Damage Probability Matrices for Prototype Buildings.* Seismic Design Decision Analysis Report No. 8, Structures Publication No. 380, MIT Press, Cambridge, Massachusetts.

Wiesner, C. J. 1964. Hydrometeorology and flood estimation. *Proceedings of the Institution of Civil Engineers,* **27:**153–167.

Wuorinen, V. 1974. A preliminary seismic microzonation of Victoria, British Columbia. Unpublished M.A. Thesis, Department of Geography, University of Victoria, Victoria, B.C.

Wuorinen, V. 1976. Seismic microzonation of Victoria: A social response to risk. *In:* H. D. Foster (Ed.), *Victoria: Physical Environment and Development.* Western Geographical Series, 12. University of Victoria, Victoria, B.C., pp. 185–219.

Wuorinen, V. 1979. A methodology for mapping total risk in urban areas. Unpublished Ph.D. Dissertation, Department of Geography, University of Victoria, Victoria, B.C.

4
Safety by Design

Building codes are designed to: establish minimum safeguards in the construction of buildings, protect occupants from fire hazards or the collapse of the structure, and prohibit unhealthy or unsanitary conditions. They deal with new as well as existing construction. These regulations are laws; they represent minimum requirements and are widely accepted as necessary for the public interest.

Working Group on Earthquake Hazards Reduction (1978)

Microzonations permit spatial variations in risk to be used as criteria for determining optimum land use. They have the potential to greatly reduce disaster losses by the avoidance of sites that have unacceptably high risk characteristics. This is particularly true when this strategy of disaster mitigation is combined with others adopted to save lives and minimize property damage. Losses can also be lessened, for example, by continuously seeking ways of avoiding risk through improvements in the design, maintenance, and use of the urban infrastructure, including its buildings and technology. Although the decisions involved will be taken normally by politicians, planners, and others outside the safety committee, these individuals can often be influenced by information and suggestions on methods of reducing risk. The safety plan coordinator and his support staff, therefore, have a responsibility to monitor changes in a wide variety of policies and to evaluate their impact on risk, pointing out any negative effects and proferring better alternatives where feasible.

Such an approach to safety is not new. The reduction of life loss through the adequate design and construction of buildings has been a major concern for centuries. As early as 1750 B.C. the Babylonian Code of Hammurabi included the following tenets (Legget, 1973):

> If a builder builds a house for a man and does not make its construction firm and the house which he has built collapses and causes the death of the owner of the house, that builder shall be put to death. . . . If it [the collapse of the house] destroys property, he [the builder] shall restore whatever it destroyed, and because he did not make the house which he

> built firm and it collapsed, he shall rebuild the house which collapsed at his own expense.

Although the legal penalties for error in building have been reduced somewhat, and the guarantees given to owners made less comprehensive, the mitigation of risk to life and property is still a major concern of the construction industry.

Change in building type, utility operation, and industrial composition can rapidly alter the risks to which a community is subjected. For this reason all change must be monitored by the safety committee. Indeed those involved in planning land use and with invention, design and construction must be cognizant of six basic aspects of safety (Adcock, 1978). These are structural integrity, operational compatibility, fail-safe design, forgiving environments, emergency strategies, and user security. As far as possible the principles involved should be incorporated into all modifications of the infrastructure in an effort to ensure that risk is reduced rather than increased by social change.

Structural Integrity

The structural integrity of a building is its physical adequacy for its intended function. Typically, major disaster areas are littered with the wreckage of buildings and utilities that have been unable to maintain their structural integrity under abnormal conditions. While architects and engineers do not, nor should not, seek to ensure absolute safety, there is little doubt that improvement in construction practices can reduce casualties and property losses. To this end, design and construction should attempt to provide structures that meet established acceptable risk levels at the lowest cost. Collapse or damage may subsequently occur, however, because of errors made during design, construction, or use.

The need to design buildings capable of withstanding the impact of disaster agents was graphically illustrated in Tangshan, China, on 28 July, 1976. When a major earthquake occurred with its epicenter directly under the city, 352 multistory buildings were seriously affected. Of these, 177 were completely destroyed, 85 suffered partial collapse, and 99 were severely damaged. Only four retained their structural integrity. In addition, 20 highway bridges collapsed and a further 211 suffered damage, as did 40 earth dams. Although no official death toll has been released, 750,000 fatalities have been estimated (Sullivan, 1979).

Rodin (1978) has identified four approaches to construction which are aimed at reducing hazards or mitigating their consequences. The first such method is the use of a probabilistic approach to design and construction. It recognizes that total strength and structural response are the end prod-

PLATE 7. Destruction in Santa Maria de Jesus, Guatemala, caused by the February 4, 1976, earthquake. Many buildings in the developing world are extremely prone to earthquake damage (American Red Cross photograph by Ted Carland).

uct of numerous statistically variable components. This approach allows the possibility that load may exceed strength to be calculated using probability theory. In consequence, risk can be expressed as a probability of failure. The aim then becomes that of providing an acceptable probability that unserviceability or collapse will not take place. This approach involves several steps including calculation of the loads likely to be experienced by a building under the influence of disaster agents, for example, earthquakes or hurricanes. To establish this, existing data on the occurrence of adverse events are often analyzed using mathematical models such as Poisson or binomial distributions to predict the return periods of disaster agents of differing magnitude (Blume, 1978). Such a probabilistic approach also requires a detailed knowledge of the strength and other characteristics of different building materials and structural types. This problem is compounded by the constant innovations occurring in materials, design, and construction practices (Knoll, 1978). Using a probabilistic approach to guaranteeing structural integrity also necessitates the implicit or explicit acceptance of certain levels of risk, since under certain extreme conditions the possibility of failure must be accepted. In this respect is it very similar to microzonation, since unless some risk were permitted no development would be allowed. It is essential to ensure that the risks involved in a particular design are clearly appreciated by the architect, engineer, and occupants.

This approach suffers from one further serious drawback as Knoll

(1978) has pointed out, "we still know practically nothing on the statistical properties of human error and their consequences, except for the very general statement that they are responsible for the majority of accidents." It is possible for errors to occur in calculation or construction which result in far greater risks being taken than those that were anticipated. As far as possible designs must therefore be carefully checked before construction permits are issued and work must be inspected while in progress.

A second approach to providing structural integrity involves a historical review of building failures and the identification of the specific factors which have caused them. In this way it is expected that proneness to failure can be predicted and therefore reduced by taking advantage of adverse experience. A safety committee can play a significant role in collecting and evaluating the literature assessing such failures. Pugsley (1973) considered the chief factors that have influenced past structural problems to be the use of new and unusual materials, difficulties stemming from the method of construction or type of structure, inadequate research and development background, inexperience and poor organization of the design and construction teams, and the industrial and political climate. If building failures during disaster are to be reduced, then data on their causes must be analyzed and used to improve procedures. To this end the causes of serious structural weakness and collapse must be studied in detail. For legal and other reasons such information is often not widely available. To overcome this difficulty, a procedure has been established in the United Kingdom for summarizing the evidence in an anonymous way, while in the United States the Engineering Performance Information Center has been established to carry out this task (Rodin, 1978). As Knoll (1978) has pointed out, it is this second approach to guaranteeing structural integrity that is most commonly in use within the engineering and architectural professions. Indeed many alterations to building codes and standards have been the result of adverse experience with existing regulations elsewhere. This approach gains in validity with the size of the data bank on structural performance in disaster.

An alternative third approach to ensuring the stability of a structure is to design it so that it will not collapse even after certain critical components have lost their effectiveness. To this end, the design is manipulated to establish alternative load pathways so that there is a broad level of protection even against unanticipated hazards. This concept was incorporated into building regulations in the United Kingdom after the Ronan Point highrise apartment building collapse (Rodin, 1978). Here a gas explosion blew out an external wall panel, triggering a "progressive collapse" of the wall panels, first above and then below the fatal apartment. The resulting inquiry discovered serious structural weaknesses in the load-bearing wall paneling of the building. It also predicted that further such "domino" accidents were more likely to result from high winds than gas explosions (Council for Science and Society, 1977). To date this safety precaution has not been included in United States building codes but preliminary criteria based on those in force in Britain are included in HUD/FHA requirements for highrise precast concrete construction. Warning-type clauses based on this concept are included in French regulations and it is in limited use in the USSR. In the Canadian Building Code, the designer is obliged to protect the user against progressive collapse by applying one of two strategies. The first, already described, involves equipping the structure with alternative paths for loads should any single member fail. The second alternative safety method is designing to accommodate the worst possible assumption, that of "the heaviest locomotive hitting the main column with full speed." In certain cases, such as with bridges, the second alternative is commonly followed and they are designed so that their main piers can withstand maximum expected ice pressure or ship impact. In the case of multistory buildings the first option, the use of alternate path for loads, is followed (Knoll, 1978).

A fourth approach to ensuring safety in construction involves analyzing the social consequences of failure. Structures are ranked according to their significance. The more disastrous the consequence of their collapse, the greater the stringency of design and inspection. The Nordic Building Regulations, for example, propose three safety classes, graded according to the implications of failure and, therefore, the degree of reliability required. In the United Kingdom such an approach has been taken in bridge and dam design, the former are graded as to span, type, and complexity and are subjected to increasingly rigorous checking procedures as a result. Large dams may only be designed by specially licensed engineers (Rodin, 1978).

The safety plan coordinator and his committee must address two distinct aspects of structural integrity in their community. The first of these involves problems associated with the exisiting infrastructure. Building codes and standards have evolved gradually. This means that in most areas there are many structures that do not conform to existing building codes, but are exempt because they predated the amendments that they now violate. In Los Angeles, for instance, the number of buildings that do not meet present-day standards for earthquake resistance has been estimated at between 20,000 and 50,000. This stock represents a relatively small proportion of the total buildings in the city because, since 1933, Los Angeles has required all new structures to be of earthquake-resistant construction. Large numbers of nonconforming buildings can be found in almost every other California community and elsewhere. The cost of reinforcing them so that they would not collapse during a strong earthquake would be billions of dollars. When considered on a national scale, "the situation becomes hopeless" (Working Group on Earthquake Hazards Reduction, 1978). Internationally the problem of inadequate structional integrity is far worse, and when all hazards are considered, it is apparently insurmountable.

Nevertheless, certain steps can be taken to improve the situation. The safety committee can commission a survey of the structural integrity of all buildings in its area. This information can then be mapped and kept in computer data bank form. It will be of value in promoting the demolition, closure in emergencies, or reinforcing of very high risk buildings. It can also be used in the preimpact phase of a disaster plan to ensure that the occupants of low integrity structures, located in high risk microzones, are evacuated first. Such data are of great value in disaster simulations of various kinds, where its role is described elsewhere in this volume. Information on structural strength and probable building performance in disaster situations is also critical in planning a community insurance program.

The problem of the threats to safety posed by older structures is too large to be handled by local communities alone. Federal and regional governments should also be encouraged to promote improvements by providing tax incentives for this purpose to private owners. Abatement ordinances can be adopted locally, however, to reduce risk from some aspects of poor building design. To illustrate, parapets and loose bricks and stones in facades above street level can fall and injure motorists and pedestrians or block emergency traffic even in minor earthquakes. Such hazards can be controlled by an abatement ordinance (Working Group on Earthquake Hazards Reduction, 1978).

There is also no guarantee that all new buildings proposed in a community will be structurally sound. In the United States, for example, there is considerable confusion over the administration of building codes. The legal authority for building regulations is vested in state governments, although in the past much of this power has been delegated to local governments. In consequence over 5000 different local codes exist. This multiplicity led to criticism which resulted in 22 states adopting some form of statewide building code. The National Conference of States on Building Codes and Standards was established in 1967, with the major objective of achieving more national uniformity in the regulation of buildings (Working Group on Earthquake Hazards Reduction, 1978). As a result of this complexity in the United States, which has parallels in Canada and elsewhere, adherence to building codes does not necessarily guarantee structural integrity in disaster. One common weakness is that such legislation often does not recognize the rapid spatial changes in risk which can occur over very short distances. Uniformity in requirements may sound equitable to developers but it does not lead to common levels of risk. In addition, there is always a considerable lag time between the discovery of new hazard-related information and the necessary resulting revision of building codes.

The role of the safety committee in this respect might be that of a catalyst. It should seek to ensure that presently adopted building codes are strictly enforced and that exceptions are not made without a full appreciation of the risks involved. An attempt might also be made to solicit building codes from other communities which face similar hazards. Any beneficial innovations these contain can then be brought to the attention of the local council for possible adoption. To illustrate, regions that suffer from cyclones would be well advised to consult the *Model Minimum Hurricane-Resistant Building Standards for the Texas Gulf Coast* published by the Texas Coastal and Marine Council (1978). Such a crosschecking of experience and building requirements may assist in ensuring safety in design. While it is important to have sufficiently stringent building codes, it must be clearly understood that if these are too restrictive they can cost more to satisfy than would be lost as a result of disaster agent impact. A balance must be achieved therefore between the benefits and costs of safety in building design.

Operational Compatibility

Numerous accidents and even some disasters have occurred because of an incompatibility between machines and their operators. There are many reasons why this problem is increasing. Technology is becoming ever more complex and, as a result, relatively high levels of intelligence and rigorous training are often required to operate and maintain it. In some cases the effects of the operation of this equipment is such that its health and environmental ramifications are not fully understood by its operators. At the same time, sophisticated machinery is being used in societies where there is no reservoir of technological expertise. Illiteracy may be endemic or instructions may be given or parts identified in a language unfamiliar to its potential operators. In addition, the widespread use of drugs and alcohol and problems of mental illness may temporarily or permanently impair an individual's ability to handle dangerous materials or sophisticated systems.

Operational incompatibility often occurs where technology or chemicals produced in the Developed World are given or sold to the Developing World. For example, on 16 September 1971, the freighter S.S. Trade Carrier unloaded 16,000 tons of wheat at the port of Basra, Iraq. This had been sprayed with a pink dye to warn that it had been treated with a deadly poisonous methyl mercury fungicide. The seed was for planting only and its sacks carried a warning to this effect in Spanish. Many farmers were unimpressed and some, having fed the grain to chickens without immediate ill effects, began using it as animal feed or for family meals. Mecury, however, does not strike immediately but accumulates in the body, eventually attacking the brain and nervous system. By late 1973 perhaps as many as 6000 had died of this poison and 100,000 may have

PLATE 8. Waves on Lake Pontchartrain, Louisiana, generated by Hurrican Hilda in 1964 caused damage to shorefront property (American Red Cross photograph).

been injured, many seriously (Hughes, 1973). Clearly it is essential that those at risk from a disaster agent must be fully informed of its potential physical consequences and of the speed with which these are likely to occur.

There are many ways in which government can attempt to reduce this problem. Educational standards must be maintained and literacy tests carried out on all those involved in potentially dangerous situations. Instruction manuals should be published in easily understandable language familiar to its intended users. Designers should attempt to identify potential areas of confusion and remove them wherever possible. The safety committee should encourage all agencies and industries in the community to ensure that the health of employees is checked regularly and alcoholics and drug users treated and removed from the workforce until cured. Microcomputers coupled with ignition switches can be used to ensure that dangerous machinery is not activated by those who, because of alcohol, illness, or other disability, are not capable of completing certain tests designed to verify their abilities. In addition, public access to dangerous weapons and machinery can be limited or prohibited by restrictions on sale or use.

Fail-Safe Design

Ideally, one of the best approaches to mitigating the consequences of deficiencies in buildings, machines or operators is the use of fail-safe principles. This basic concept involves the design of a system so that it functions with negative (self-correcting) rather than positive (accelerating) feedback. In this way systems can be built which, even if they fail to function as expected, will do so in a manner that is not dangerous. Large transport trucks illustrate this principle (Adcock, 1978). These are normally equipped with brakes that are held off by air pressure. If this pressure is lost because of some malfunction, the brakes automatically go on, stopping the vehicle and so reducing the possibility of disaster. Traffic lights are normally designed to turn to red should a failure occur. In this way all traffic is obliged to stop, preventing accidents.

The safety committee should actively attempt to make sure that, as far as possible, buildings and equipment used in their community operate on this principle. Although it is very difficult to apply at an urban infrastructural scale, it can often be used to prevent small threats from developing into larger ones. The equipping of buildings with sprinkler systems which are activated by heat is an example of its increasing application. Design approaches to stopping fire spreading are illustrated in Figures 4.1 and 4.2.

Near Frankfurt, West Germany, 100 cars have been fitted with radio devices designed to summon aid at the press of a button. AEG-Telefunken has developed the system which gives the driver a choice of three buttons: one for breakdowns, one for accidents, and another for voice messages. If the breakdown or accident button is pressed a signal is transmitted to the nearest relay station and from there to an emergency center. The vehicle is located automatically by radio direction finding equipment and help dispatched (Canadian Press, 1978).

One important step in increasing the use of fail-safe design has been the invention of "bleeding" machines. Hollow bolts are now being manufactured which are filled with a small amount of vivid fluorescent dye. When the bolt begins to crack, for example, from metal fatigue, the dye is spread by capillary action. A liquid warning is therefore provided on the surface of the structure, whether an aircraft, automobile, or virtually any machine containing moving parts. This concept need not be limited to bolts but can be applied to any element locking two parts together, such as the shafts holding helicopter rotors, drydock gate hinges, or rods in building cranes.

It may prove to be difficult to ensure that such fail-safe design principles are widely adopted. Nevertheless, the safety committee should, at a minimum, try to assess every development to determine whether it might not lead to the reverse, a "domino effect." Under certain circumstances, for example, a small problem may be magnified by poor design so that it can spread or diversify rapidly, leading to catastrophic failure.

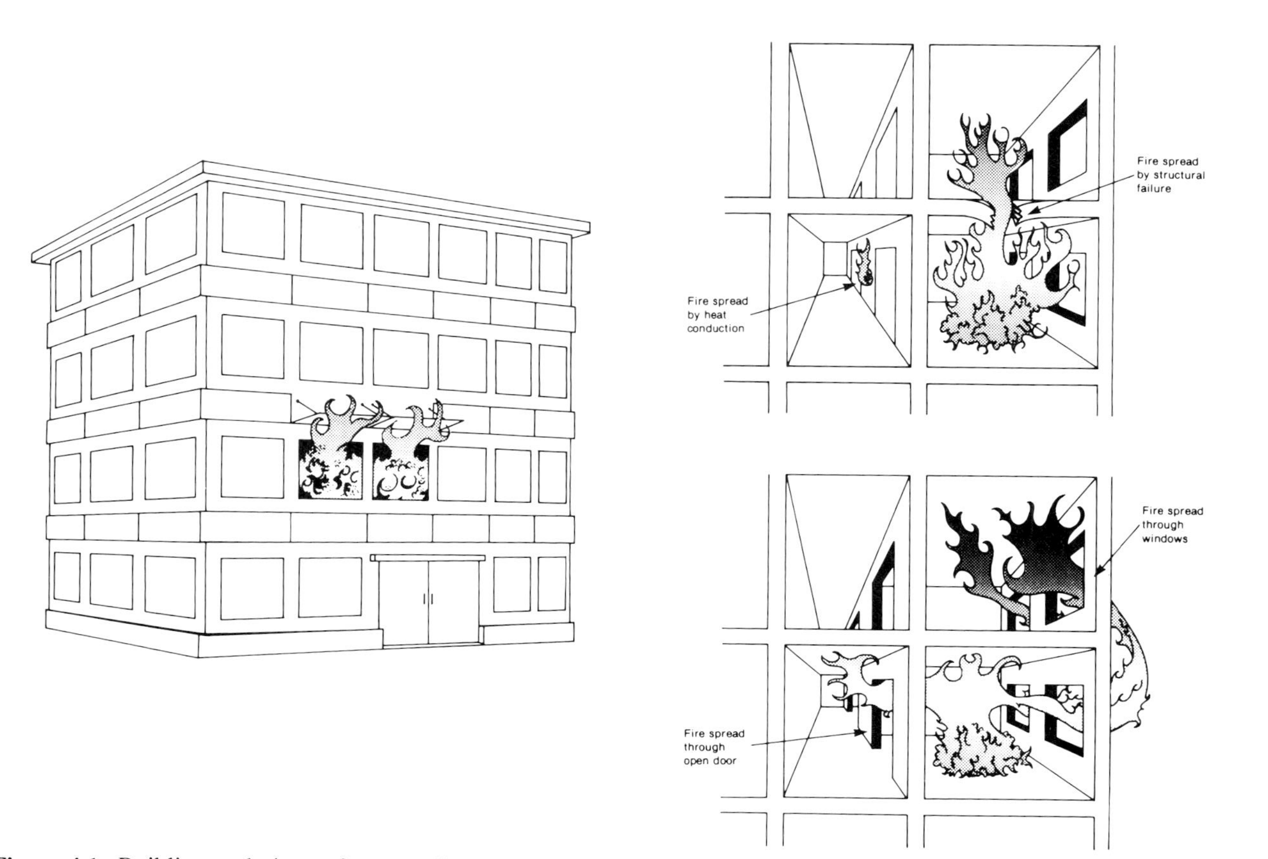

Figure 4.1. Building techniques for retarding the spread of fire, flame deflectors in operation (T.Z. Harmathy, 1977).

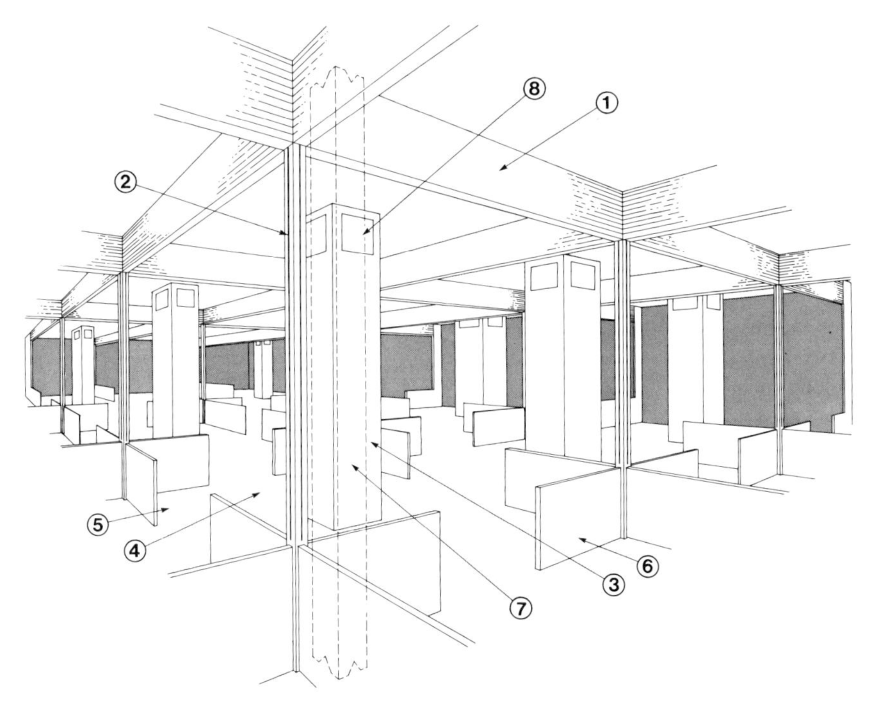

Figure 4.2. Internal design and fire retardation. The figure shows a large uncompartmented space equipped with a fire drainage system. The ceiling is divided into many rectangular areas by a series of retracted fold-up drop curtains (1) made of light-gauge metal and equipped with weightier bottom pieces. The purpose of these curtains is twofold: they restrict the spread of flames and smoke during the growth period of fire; and when activated by the fire, they slide down in grooves (2) to floor skirting boards (3) and surround the cell on fire (4), leaving only four openings (5) properly sized for controlled ventilation.

There is a column (6) in the center of each cell. A well-insulated "drainage duct" (7) runs the entire height of the building in the interior of the columns. Each duct has four "access gates" (8) (insulated on the duct side) near the ceiling on every story, by which it serves a number of cells located on the successive stories. These gates are normally closed by simple fusible parts. There are two or four "release gates" (not shown) at the top end of each drainage duct above the roof level. They are held closed by the tension of a heat-destructible line extending to the bottom of the duct.

As fire in the cell starts to build up, the access gates (8) open shortly before the activation of the drop curtains (1). The fire gases enter the drainage duct (7), and by destroying the tensioning line cause the release gates at the top to open. Not only are the gases and flames safely withdrawn from the building, but the suction created by the column of hot combustion products in the duct creates a depression in the fire cell and thus prevents the dispersion of smoke and fire to the neighboring spaces (T.Z. Harmathy, 1977).

Forgiving Environment for Failures

A forgiving environment for failures is created when planners, architects, and designers assume a relatively high incidence of destruction or misuse and try to minimize its consequences. It is possible to contemplate applying this principle at a variety of different levels. For example, when designing a land use policy, care must be taken to ensure that should the destruction of one component, such as a chemical factory or oil refinery, occur, this does not cause secondary disasters among adjacent land uses. To provide such a degree of safety, disaster simulations are essential. These should allow the definition of zones of risk around major noxious facilities in which other development is strictly controlled. This is rarely the case. In their survey of London, Ontario, Hewitt and Burton (1971) determined that at least 50% of all industrial premises were situated too near to their neighbors, perhaps making them susceptible to chain reactions in disaster situations.

Forgiving environments can be created within buildings (Pauls, 1975). Unfortunately with the trend toward highrise development this is frequently not the case. In a highrise building in Sao Paulo, plastics fed a fire that in 25 minutes engulfed the top 14 floors of a 22-story office and garage building killing at least 189 people. Poor design also contributed to this disaster. The building had four elevators in a central core. These jammed during the fire, cremating an unknown number of victims. The building had no automatic sprinkler system, no interior fire stairs, and no exterior fire escape. Less than two years earlier, in April 1972, the same city saw the destruction of a 30-story building in 15 minutes by a similar fire. Plastics were also responsible for the disastrous Summerland fire in the Isle of Man. Summerland was the largest building on the island and the biggest entertainment center in Europe. A plastic hut on a minigolf course outside the building was set alight. It collapsed against the southeast wall made of corrugated metal sheeting called Galbestos which was coated with a material that gave off highly flammable gases when on fire. Within a minute flames had spread within the whole cavity igniting most of the wall. As a result 50 people died, nine of them children, in the greatest fire-related British life loss since World War II (Hutcheon, 1974).

Numerous attempts have been made to overcome the evacuation problem posed by highrise buildings. In England, the City of London police force has undertaken a survey of the 105 highrise buildings that do not possess a ready means of escape to adjacent roofs. The height of each building was recorded as was its topography and whether it would support the weight of a helicopter. Details of any obstruction over 1 ft in height were noted as was the presence of unfixed equipment or loose gravel chippings. In summary, of the 105 buildings, 66 could easily accommodate a helicopter, 25 could possibly permit one to land on a restricted area, and 10 were so constructed or obstructed that landing would be impossible. Four buildings were under construction and their architects agreed to ensure suitable areas for helicopters to land (Fisher, 1975).

In the case of an emergency, the Helicopter Rescue Scheme is put into operation. The Air Traffic Control Centre directs a suitable helicopter to rendezvous with disaster related personnel and establishes a "prohibited zone" over the incident area that can only be entered by helicopers engaged in rescue operations. The helicopter pilot is briefed, given a plan of the affected building's rooftop and flies emergency personnel to the scene, evacuating trapped victims in an orderly fashion.

To facilitate such an operation the London police suggest a minimum area of 50 ft by 50 ft, illuminated by quartz–halogen white lights be kept free of obstructions. This must be capable of withstanding 30 pounds/sq ft impact load. All obstructions over 1 ft in height should be painted in fluorescent red and orange and where feasible these should be hinged so they can be lowered in an emergency. It was also proposed that an identifying code number be painted on each rooftop to guide helicopters to the building (Fisher, 1975). It is suggested that all safety committees conduct a similar highrise survey, establish emergency procedures, and make arrangements with civil operators and the armed forces to supply helicopters for emergency evacuations. The provision of suitable rooftop landing facilities could be made a criterion for the granting of building permits for highrise structures.

Unfortunately, updrafts from major fires may make helicopter evacuation infeasible. This happened in October 1977 when fire raged through a modern nine-story office building in Lima, Peru. Three hundred firemen and a helicopter were unable to help those trapped above the fire. At least six occupants were killed by either fumes or by jumping, while 12 more office workers suffered broken bones and concussions leaping from lower floors. Adequate building design can overcome this problem. A recently invented concept consists of a series of hinged balcony floors which can be lowered by means of a lever. This produces a substitute stairway on the outside of the building which is far less likely to be blocked by smoke or fire. Its cost, as a percentage of that of the building as a whole, is virtually negligible.

An alternative means of escape has been invented in France. This consists of a long sausagelike tube of synthetic elastic material which hangs from a strong metal hoop attached to the building and can be lowered to the ground along the outside in case of fire or other emergency. This is approximately the width of a human body and victims ease themselves feet first into it and slide down. Their rate of descent is some 6 ft/sec but can be reduced by pushing their elbows against the tube. Under test conditions 35 people/min can be evacuated, regardless of their size and weight.

Such systems are relatively cheap and could easily be made mandatory in highrises. Mention should also be made of the fire safety program being sponsored by the U.S. National Association of Insurance Agents. This involves providing special high-visibility decals to mark the windows of rooms occupied by invalids, children, and old people who are unable to evacuate under such circumstances without assistance.

An increasing number of disasters can be traced to the unwillingness of architects and designers to create forgiving environments. The use of synthetic materials that generate noxious gases during combustion or of glass in highrise buildings subject to earthquakes are obvious examples. This problem was graphically illustrated by the crash of a six-car train at Moorgate Underground Station, London, on 28 February 1975 (Fisher, 1976). The train, traveling at some 40 miles/hour across its normal stopping point, continued through a sand bag, demolished a hydraulic buffer stop, left the track, and struck the solid wall at the end of a short, blind tunnel. As a result, a total length of 118 ft of train became compressed within a 66 ft tunnel, causing 43 fatalities in the process. With better design, including a longer tunnel, more and graduated resistances and an automatically activated braking system, the magnitude of the impact and hence the casualties could probably have been greatly reduced.

One aspect of the creation of forgiving environments which requires far greater consideration is that associated with the transportation of dangerous substances. Prior to 1900, most trade goods were inert. Today some 1700 hazardous materials are regularly in transit through built-up areas in the Developed World (Rodin, 1978). In Britain this represents some 10% of all transported substances. These materials may be solids, liquids, or gases, often under pressure or at low temperature, that can be extremely dangerous when subjected to fire, accidental impact from collision, or equipment malfunction. In Britain alone 40 to 50 million tons of such materials are transported annually, resulting in between 200 to 300 incidents each year which, at a minimum, cause road closures and traffic delay.

The British experience has been relatively reassuring, 7 deaths in 44 accidents in the past 10 years (Rodin, 1978). This has not necessarily been the case elsewhere. In 1977, for example, accidents involving dangerous goods in transit caused 32 deaths and 543 injuries in the United States. The most hazardous materials proved to be gasoline and liquefied petroleum gas, responsible for 26 deaths and 102 injuries, and anhydrous ammonia and chlorine which caused 5 fatalities and 125 injuries (Johnstone, 1978). The problem continues; in February 1978 a serious accident occurred at Youngstown, Florida where a tank car containing chlorine was derailed and ruptured. Some 12 people died and 40 were hospitalized as a result, while approximately 1000 local residents were evacuated. In this instance the problem was compounded since many of the victims were driving along U.S. 231, a highway which parallels the railroad tracks where the accident took place. The gas caused car engines to stall, trapping and asphyxiating the occupants.

All governments should be involved in keeping such incidents to a minimum. At the federal level in Canada, for example, the Transportation of Dangerous Goods Branch has been established within Transport Canada. Its chief objectives are to promote safety, standardize regulations, and coordinate the activities of all levels of governments in dealing with this problem. The Branch has undertaken four main activities to meet these objectives. The first has been the preparation of a single national reference document, the "Transportation of Dangerous Goods Code" in which manufacturers, shippers, and carriers can find all the information required to safely move hazardous materials by any transportation mode. A second draft of this document was published in 1978 and circulated to various manufacturing and industrial bodies and to government agencies for their comment (Johnstone, 1978). It includes a single danger-classification system and nomenclature, uniform documentation, labeling and placarding regulations, and the development of a package certification program which is based on performance standards. The classification of such hazardous materials, the labels and placards used to identify them, and the standards to control their packaging are largely those recommended by the United Nations Committee of Experts in the Transport of Dangerous Goods. This helps to ensure that Canadian federal and international standards are compatible, so facilitating the export trade. The situation in Canada contrasts with that in many other countries such as the United States where there is a massive set of regulations that cannot be amended without prolonged protests by those involved.

In addition the federal government of Canada is establishing a 24-hour emergency response and information center. A duty officer will be supported by scientific, legal, and technical specialists and staff to provide detailed information on how to deal with hazardous materials under any circumstances. Inspection programs will be greatly expanded including those for packaging manufacturers and shippers. This will be possible because a program is also contemplated to provide indepth hazardous materials training to certain key inspectors and preliminary education to general inspectors employed for other purposes and to police and fire brigade personnel (Johnstone, 1978).

Risks from such materials can also be reduced by requiring drivers to meet stringent proficiency levels and ensuring that they work relatively restricted hours to prevent fatigue-related accidents. In addition, clear vehicle marking can reduce risk by providing warning and when an accident has occurred by indicating what it contains and how the danger can be mitigated. Safety coordinators should make sure that city disaster plans include the provision for effective action in the case of spills or derailments. The necessary protective clothing, equipment, and chemicals must be readily available to permit an optimum response. In British Columbia, for example, kits costing $3500 each, containing safety gear, nonsparking tools, specialized couplings, valves, fittings, and flexible reinforced hoses to transfer liquids from damaged rail cars are being stockpiled in strategically located centers. An emergency transportation system has been set up to move these kits and experts capable of using them to a disaster scene with the minimum of delay. There are many other ways in which municipalities can mitigate the dangers caused by the transportation of hazardous materials through their jurisdictions. Risks may be reduced by planning which attempts to locate industrial users of dangerous materials in areas where they can be served by transportation routes which do not pass through city centers, tunnels, or other confined spaces. Such industrial activities should also be located in relatively low total risk microzones because of the secondary disasters that can occur if storage tanks are ruptured or explode. It may be prudent to prohibit the transport of hazardous materials during certain times of day, as for example morning and evening rush hours.

A useful safety option open to many muncipalities is the designation of certain transport routes for the movement of high risk goods. Such a system is designed to ensure that the smallest number of individuals are exposed to noxious materials for the shortest length of time. It is normal for such routing systems to favor dual carriageway roads for access in case of accident. Some emphasize the use of the rail network because of its more impressive accident record, but lines often pass through densely settled areas while trunk road systems rarely do. However, major highways often carry a large volume of traffic and an explosion or inferno can cause many fatalities among other road users. Every safety committee should consider all transporation routes on the basis of local merit. A voluntary routing system is currently in operation in Cleveland County, while a compulsory scheme is in force in the Netherlands (Rodin, 1978). It is also prudent to exercise extreme care over the location of pipelines and ultrahigh voltage power lines. The former often carry large quantites of dangerous materials, while the latter are the subject of an ongoing controversy over the effect of high tension electrical fields on human and animal health. The threat posed by pipelines is illustrated by a recent Canadian experience. Almost 20,000 people were evacuated from their homes in the southeastern Edmonton suburbs of Millwoods, Blue Quill, and Kaskitayo on 2 March 1979 when propane and butane leaked from a ruptured pipeline into the area's sewer system. Because of the low temperature the gas remained liquefied and spread over a large area. When air temperature rose it was ignited by the heat from the engine of a passing truck. The burning gas created a wall of fire and threatened over 40 residential blocks.

The possibility of improving the safety record of noxious materials carriers is increasing. With the development of computer terminals and telemetry equipment suitable for installation in vehicles, it is becoming feasible to monitor all movement. Such a federal or state network of observance would allow government authorities to know what hazardous substances were in transit, where they were located, how fast they were being transported, and along which routes. Driver hours could also be more easily controlled.

While the land and air carriage of an increasing complex array of potentially lethal goods constitutes an increasing risk, much greater hazards are often associated with marine transport of such materials. This stems from the fact that far larger quantities are moved in a single ship. Marine disasters have a long history . The world's worst accident occurred when a French munitions ship caught fire and exploded in Halifax harbor on 6 December 1917. The blast destroyed a square mile portion of the north of the city, killing at least 1963 people and injuring many more. Until very recently compensation was still being paid by the Halifax Relief Commission to surviving victims, many of whom were blinded by the hail of flying glass caused by the blast. The risk of even greater marine related incidents is increasing with the movement of large quantities of liquefied gas and other noxious materials on a global scale.

While oil tankers are less prone to explosions, they are spilling their contents with unfortunate frequency along the world's coastlines. Recent examples include the wrecking of the Amoco Cadiz on the Brest coast of France and of the Metula in the Strait of Magellan a mile from the coast of Chile. The former tanker polluted approximately 40 miles of beaches and caused losses in excess of $25 million. The distribution and persistence of oil residues on any shoreline after a major spill will reflect the volume of oil lost, the climatic, meteorological, and oceanographic conditions, the nature of sediments, and the energy characteristics of the littoral zone. Ideally, coastlines that are at high risk from tanker traffic should be surveyed and contingency plans made for their cleanup. Safety committees facing this potential problem are referred to one such case study, *Coastal Environments, Oil Spills and Clean-up Programs in the Bay of Fundy* (Owens, Leslie, and associates, 1977). In addition, they may wish to consult *An Oil Spill Bibliography* (Fingas and Ross, 1977), which provides hundreds of references on such topics as biodegradation and the physical and chemical properties of oil and its effect on biota.

Among the prudent measures that federal governments can take to reduce potential losses are establishing contingency funds, supported by taxes derived from the companies involved. These would be used to pay for cleanup costs and to compensate fishermen, resort owners, and others adversely affected by oil spills. In addition, retaining booms, detergents, and other related equipment should be stockpiled at strategic points along the coast. The policing of tanker movement through relatively safe shipping lanes must be strictly enforced. Coastal facilities for tankers and their associated storage tanks must be located in low risk zones which are not in close proximity to other industrial development or residential areas. Unfortunately such precautions are often neglected. Valdez, the terminus for the Alaskan oil pipeline from Prudhoe Bay is a high risk site threatened by both icebergs and tsunamis.

One area in which consideration of the creation of a forgiving environment is essential is in the design of grandstands, arenas, and theatres. There have been a total of 125 deaths and over 1000 injuries in eight crowd-related incidents during this century in Britain. As a result, new regulations and standards have been introduced to create a more forgiving environment in all new and existing sports grounds (SCICON, 1972). Similarly, the pressurization of stair shafts as a means of providing smoke-free escape routes during fires has received much attention in recent years by researchers and code authorities (Tamura, 1974). This entails injecting air into the stair shaft to establish flow from it to adjacent space. This prevents the entry of smoke into it and disperses any already there. Studies have shown this as an effective means of creating a forgiving environment in tall buildings. Air appears to be best injected at several levels (Tamura, 1974).

In summary, safety plan coordinators and their support staff should be pessimistic. They should assume that noxious substances will be spilt in their community, that fires will break out in highrise buildings and that floods, earthquakes, or other natural hazards will strike. Every step should then be taken to ensure that even if these events do occur, the resulting loss of life and damage to property is minimized by a forgiving community environment.

Emergency Evacuation

Buildings must seek to accommodate both the commonplace and the exceptional. On relatively rare occasions fires, explosions, bomb threats, inclement weather, or earthquakes may necessitate rapid evacuation. This need poses both design and management problems. Although the evacuation of structures has been a major concern of building codes for over 50 years, its scientific and technological base is insecure. Pauls (1977) , for example, has pointed out that many building codes contain errors in their regulations governing the width of exit stairs and riser tread geometry. The latter is often based on a simple design rule developed over 300 years ago when body dimensions, stride length, and even the units of measurement were different from those today. Pauls claims that, as a result, exit stair treads are normally about two inches too shallow, a deficiency with considerable time consequences in evacuation.

Typically the so-called hydraulic model has been used by those drafting code requirements for means of egress. In simple form it envisages a building as a reservoir of water with exit routes being pipes with valves controlling drainage through the base. Evacuation performance or rate of discharge is related to the easily measured orifice dimensions. Pauls (1977) argued that this model is too simplistic, since it neglects decision-making complexities. In addition it draws attention away from the "threat" or cause of the evacuation and emphasizes the exits or safe ends of the process. In an effort to develop more realistic code requirements, the Division of Building Research of Canada's National Reasearch Council has been monitoring evacuation drills from office buildings ranging from 8 to 29 stories in height. In addition it has been studying crowd flow in theatres, arenas, grandstands, and transit facilities. Several thousand man-hours, for example, were spent in research and videotaping and filming pedestrian movement during the Olympics in Montreal.

Considerable information on how people actually behave during evacuation is now available as a consequence of these studies. For example, major misconceptions have been discovered regarding the commonly used "unit exit width" basis for determining stair width. This is based on the oversimplified assumption that people will walk two abreast down a 44 inch exit stair, which as a result has generally been adopted by code writers as a standard minimum (Pauls, 1977). An examination of actual crowd behavior shows that this rationale is invalid and results in an overestimate of the speed of building evacuation. Figure 4.3, developed by Pauls, is an attempt to summarize the times needed for various evacuation procedures using stairs, refuges, and elevators in tall office buildings. It shows that actual evacuation time depends very much on certain emergency management procedures. The structure being modeled is a multistory office building with two 44-inch wide exit stairs and four groups of 4 elevators serving the ground floors and floors 2 to 12, 12 to 22, 22 to 32, and 32 to 41. These travel at speeds ranging from 800 to 1200 ft/min and service 4500 people. In summary, total voluntary evacuation by stairs will probably take considerably longer than managed evacuation that includes the use of refuge floors and elevators. In disaster situations it may often make more sense to evacuate certain floors only, since these are at higher risk than the remainder of the building but will not be given priority in evacuation under unsupervised conditions.

It has also been determined that under normal circumstances about 3% of those usually present in highrise office buildings cannot or should not attempt to evacuate by means of crowded stairways. These include individuals with obvious physical disabilities, or who suffer from heart disorders, or are convalescents from recent illness, accidents, or surgery. This problem becomes even more acute in hospitals, homes for the elderly, and some other institutions. Every care must be taken in the design of such buildings that they can in fact be evacuated rapidly. Stairways, for example, must be capable of accommodating both wheelchairs and stretchers when elevators are out of service or too dangerous

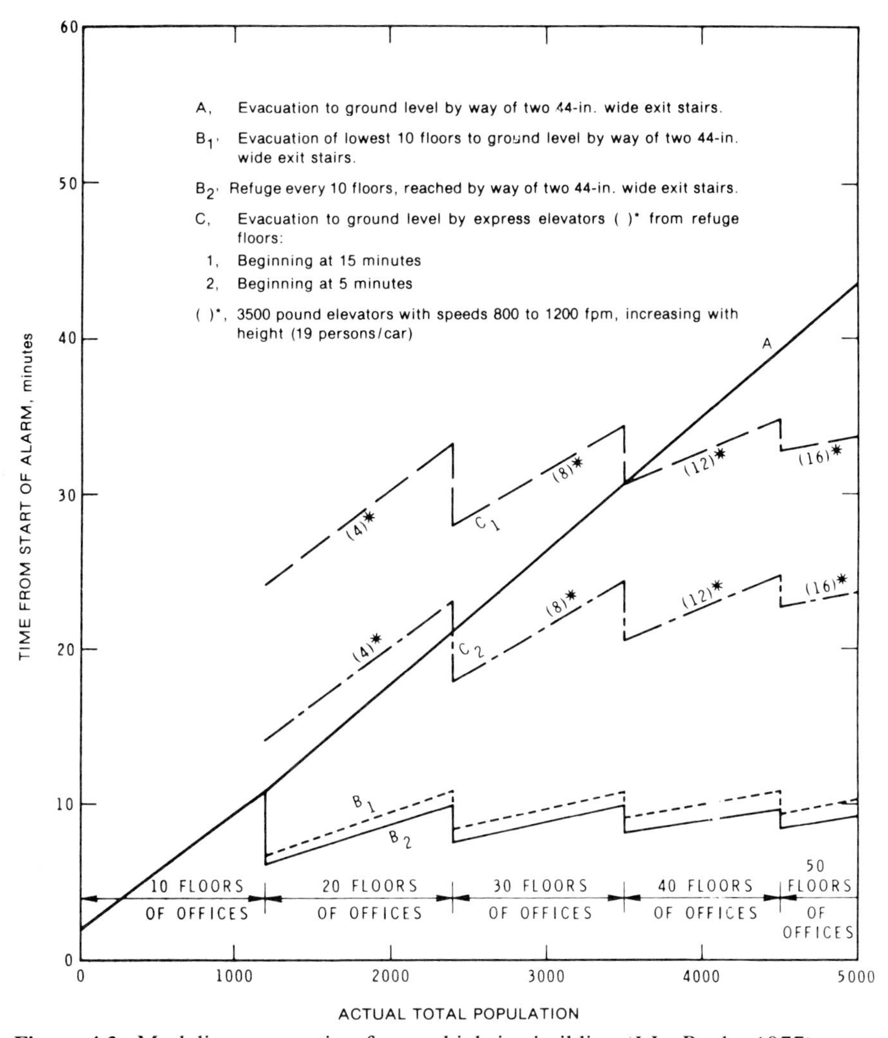

Figure 4.3. Modeling evacuation from a highrise building (J.L. Pauls, 1977).

to use. Special evacuation procedures must be developed and tested to move the incapacitated with a minimum of delay.

In summary, it has been shown that many of the emergency facilities demanded by building codes are too optimistic in their rated capacity. Safety committees would be well advised to monitor behavioral research to ensure that their codes are in fact adequate. There are many ways to evacuate large buildings. Each differs in the time taken and in its attendant risk. It should be stressed that the aim of evacuation is not to empty a building of its occupants in the minimum time but rather with the fewest casualties. To this end designers should consider the use of especially safe refuges within the building. Managers should also remember that ideally those in the greatest danger should be evacuated first. To make sure that all occupants respond as expected, emergency drills should be carried out at regular intervals to ensure that those involved are familiar with all safety procedures.

Security

It is easier to destroy than to create. Yet, unfortunately, the act of destruction is often associated with substantial benefits for those responsible. Such advantages may include publicity, prestige, economic gain, and increased power. In consequence, deliberate damaging acts against people and property are occurring with increasing regularity. This spectrum of destruction stretches from vandalism, arson and sabotage to riot, civil war, and invasion. Security in design, aimed at preventing or at least reducing such damage, is becoming of ever-increasing significance.

This is not a new concept, having given rise to such diverse structures as the Great Wall of China and the Tower of London. The moat, the portcullis, and the turret are all historically significant building innovations designed to increase security. What is new is the scale and diversity of possible acts of aggression and therefore the variety of threats that society must seek to guard against. In this respect the sheer scale of operations in modern transportation networks, energy grids, and other supply systems magnifies the effects of what are in themselves only minor acts of vandalism or carelessness. Ward (1973), for example, cites several instances of this phenomenon from Britain. These include the opening of the taps on an oil tank outside a Worcester printing works causing a 12-mile oil slick from the source to Upton-on-Severn which threatened the water intake of Gloucester and Coventry. Three boys brought near disaster to 6000 homes in the Watford area by altering the setting of a gas pressure regulator, while others appear to have caused a massive leak of explosive naphtha which threatened the lives of thousands in Liverpool and closed the River Mersey to shipping.

When acts of violence become more premeditated and sophisticated, their impacts naturally increase. In the United States alone, fire is responsible for at least $11 billion in damage, 12,000 fatalities, and 300,000 injuries a year (United States National Commission on Fire Prevention and Control, 1973). Many of these losses are due to arson prompted by insurance fraud. A lesser number stem from sabotage for social or political reasons.

The process of organized violence culminates in open warfare. Two of the three most destructive events ever to influence mankind were the First and Second World Wars. Lives lost and cities destroyed by armed conflict have probably greatly outweighed damage associated with natural hazards, particularly in the twentieth century. This phenomenon is not new, however. Between 1100 and 1800 A.D. some 42 war-devastated cities failed to recover while many that did took a very long time to do so (Chandler and Fox, 1974).

Many design and construction strategies can be used to reduce the risks of vandalism and other forms of deliberate destruction. It is possible, for example, to include the probable incidence of such damage in total risk microzonations and to locate vulnerable properties in areas where it is least likely to occur. In addition, if willful damage and its ramifications are to be minimized, then the architect and engineer must take its possibility into account during design, in the same way that they seek to minimize weathering, wear, decay, or corrosion. Miller (1973) termed this concept "building self-defence." In his study of damage caused by vandalism he was able to establish obvious links between design and types of materials used in construction and subsequent losses. For example, the use of light colored soft wall finishes, large glass panes, felt, asbestos cement roofs, or copper piping was felt to invite vandalism. Open spaces in design, such as those under stairs or between buildings and walls, were to be avoided.

Another alternative for buildings that are high security risks or of special significance for critical operations is to construct them underground. This is of course not a new concept. Petra, located in the desert part of Jordan, is an ancient underground city cut out of red sandstone. It was certainly in use by the sixth century B.C. (Legget, 1973). Wherever there are unused mines or underground quarries, conveniently located in or adjacent to settlements, these can be used for evacuation or storage during times of hostilites. During the Second World War, for example, the salt mines at Hungen, north of Frankfurt, were used to store German art treasures. Many of the most valuable exhibits from the British Museum were protected in a disused tunnel near Aberystwyth, Wales, while the Elgin Marbles were kept in the London underground system.

Safety committees would be well advised to undertake an inventory of available underground space in their own jurisdictions which would be suitable to house emergency facilities or even for industrial or residential use. One of the most significant areas of such safe subsurface space in the world occurs around Kansas City, Missouri. Here Bethany Falls limestone has been mined for a century. It has been used as building stone; for making lime, cement, and concrete aggregate; and for road construction. Large pillars of rock have been left for roof support and the total thickness of the limestone, 20–25 ft, has resulted in space which is virtually ideal for underground operations. It has been estimated that there are now some 2870 acres of mined space suitable for development beneath the area around Kansas City, with 50.6 million ft of finished space available for warehousing and other uses (Legget, 1973). This area is increasing by some 10 acres each year. Originally the miners left randomly spaced pillars to support the roof but these are now regularly placed

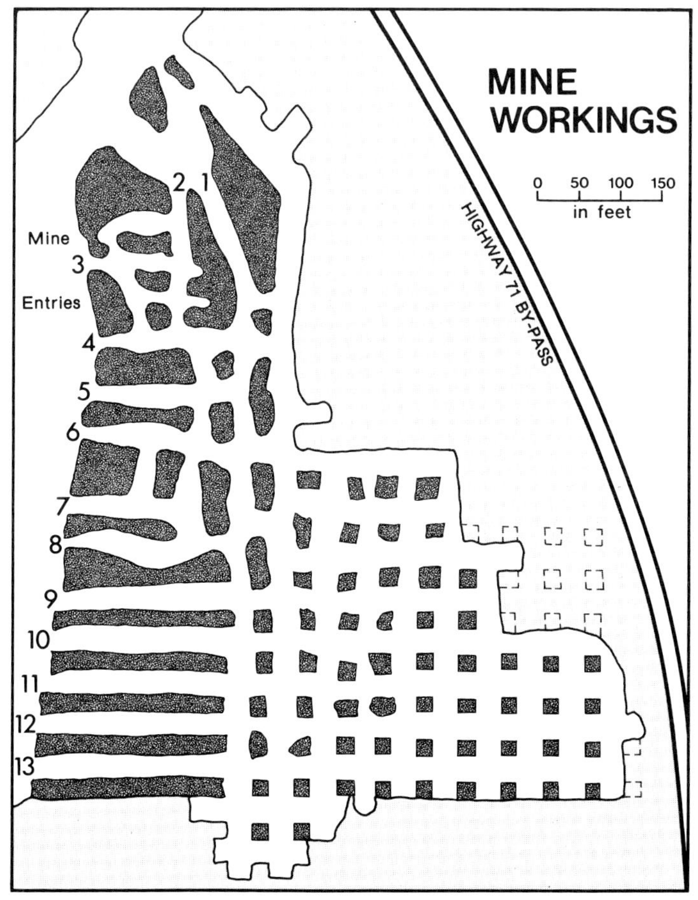

Figure 4.4. The development of underground space, Kansas City, Missouri (Missouri Division of Geological Survey and Water Resources).

with a view to later use (Figure 4.4). Such a strategy could be encouraged elsewhere.

In other areas, underground space is being created specifically as a disaster mitigation strategy. The revised Swedish Civil Defence Law of 1 July 1960 specified that all communities of over 5000 people must be equipped with properly designed "normal shelters." In case of emergency, 90% of the population of the larger Swedish cities is to be evacuated to shelters elsewhere. The remaining 10%, essential to keep these centers functioning, are to be accommodated in deep rock shelters. More than 100 of these have already been built. Even underground marine docks capable of handling destroyers have been constructed (Legget, 1973). Similarly the Command Combat Operations Center of the

North American Air Defense Command has been built as a fully equipped underground installation capable of accommodating a staff of 700. It is located near Colorado Springs, Colorado, where it was excavated almost entirely into the Pike's Peak granite in Cheyenne Mountain.

Underground facilities are also being constructed to store petroleum products and other vital fuels and chemicals. Naval fuel is kept in 20 cylindrical vaults excavated into lava at Pearl Harbor, Hawaii. These are almost 100 ft in diameter and 250 ft high. Japan is developing a series of such storage facilities to mitigate the impact of artificially created oil shortages.

Safety committees might also consider promoting the use of underground housing. It is resistant to wind, frost, and fire damage, inaccessible to arsonists and vandals, and less prone to enemy attack. It is also very energy efficient and possibly cheaper to construct than conventional homes. They should certainly encourage local governments to remove any bylaws or zoning restrictions which inhibit its use.

Paralleling the increase in the potential for deliberate destruction has been a rise in the diversity of security equipment. Many of these devices function by combining sensors of various kinds with communications systems. Intruders, fires, or other unwanted effects can be detected and the authorities notified automatically of their presence or occurrence. Such systems are now becoming available even at the personal level. Protectalert, for instance, is an electronic device worn around the neck on a pendant or in the pocket, which includes a miniature transmitter (Davy, 1979). If its owner is attacked or taken ill a slight squeeze will activate it. The transmitter then signals an automatic telephone dialer to ring a series of preselected numbers until one is answered. A prerecorded message is then broadcast giving the location of the victim and requesting assistance. Protectalert also includes a smoke detector. If a fire begins in close proximity, a relay activated by the detector's sirens would then telephone the fire department and a recorded message would give the location.

A new telephone–computer network is currently being tested in Calgary, Alberta, in 120 homes. The system has computerized medical and police assistance monitors, automatic fire and burglar sensors, remote utility reading devices, together with energy consumption and temperature monitors. Each house is connected to an emergency computer control center through the telephone network and an in-house push-button panel. When the system is activated, the computer analyzes the information and automatically dispatches the required assistance: ambulance, fire engine, or police. Such emergency crews will be immediately supplied with data on the type of home, location, number of residents, and any special medical histories. Each home is being equipped with a two-way information retrieval system that operates through the television set (Canadian Press, 1979b).

Developments are also being made with tagging explosives to counter violent crime. Taggants are added for two reasons: to aid in detection and identification. Detection taggants possess a distinctive property, such as an easily identified vapor which can be rapidly recognized by special instruments. They assist in locating hidden weapons or bombs in buildings, vehicles, or luggage. Identification taggants are microparticles included in the explosives so that after detonation their code allows the last legitimate owner to be identified. As an illustration, the 3M Company has developed a color-coded plastic chip, manufactured in pigmented layers or lamination. The 10 color layers represent the digits 0–9 and can be arranged in any order to give over one million permutations for coding and identification. Chips are designed to be fluorescent in ultraviolent light and are magnetic so that they can be easily recovered. Tests show that with the aid of a 100× microscope a tentative reading of the code can be made within minutes of an explosion and their purchaser traced within hours (Ansell, 1979).

Such devices are designed to minimize the delay between problem and adequate response. Other security-related devices are specifically meant to prevent occurrence. Many of these were originally invented for military purposes. During the Vietnam conflict the United States Defense Communications Planning Group developed an entire generation of new weapons based on sensors and remote controlled response systems. Elements in this advanced military technology included seismic sensors, intended to detect the disturbance in the ground caused by movement; electromagnetic intrusion detectors which set up a radio field around themselves and indicated when movement broke it; and magnetic sensors which reacted to the presence of certain metals. Infrared heat sensors were also developed. Such detectors were monitored in the field by computers, which themselves were often linked to weapons systems. They were used, for example, in the defense of Khe Sanh and in Project Igloo White, designed to cut traffic on the Ho Chi Minh Trail in Laos. In addition, starlight scopes, capable of magnifying existing starlight or moonlight up to 50,000 times were developed to detect nighttime troop movements. Aircraft were also outfitted with special low-light level television cameras for identifying targets in the dark (Dickson, 1976).

Although not all of this equipment is available for nonmilitary operations, some of it is now being used for security purposes. Army starlight scopes have been used by the U.S. Park Service to catch alligator poachers in the Everglades, who typically work at night. The U.S. Environmental Protection Agency is using heat sensing cameras to identify sources of water pollution. Los Angeles County firemen are using night-vision goggles as they direct low visibility forest fire fighting operations from helicopters. Such devices also have considerable potential for smuggling, vandal, sabotage, and theft control. They have been used by a Delaware telephone company to protect against pilfering of copper parts, in home security applications in a Maryland suburb, and by the U.S. Customs and Drug Enforcement Administration officials to monitor deserted airstrips. They are thought to be in use on the White House grounds and could be involved in maintaining security in nuclear power plants, warehouses, government installations, and chemical storage depots. Their largest nonmilitary use to date has been along portions of the U.S.–Mexican border. Here acoustic sensors, buried strain-sensitive cables, and infrared detection devices are in place to reduce illegal traffic (Dickson, 1976).

The use of such equipment to monitor high risk areas brings the right to safety and the right to privacy into direct conflict. This problem is growing. A 2-square-mile area on Cleveland's East Side is already covered by low-light level color television cameras that send their images directly to police headquarters by laser beam. Similarly, a section of Mount Vernon, New York, is being monitored on a 24-hour closed circuit television system set up with the aid of a $47,000 grant from the U.S. Justice Department. It seems inevitable that the civilian use of such equipment will be controlled by law to ensure that maximum safety is obtained for the minimum loss of privacy. Until this problem has been resolved at national levels, safety committees might consider the adoption of some of these techniques to monitor areas where vandalism or arson might seriously threaten their communities. One option that may be of value has been developed by Security Information Systems of Toronto. Known as Identimat 2000, this computerized device only opens doors if the shape and markings on an individual's hand match those previously fed into it by a plastic identification card. It is designed to limit access to restricted areas and to classified computer data banks (Canadian Press, 1979a). It costs $6000 per unit.

Conclusions

Many industries, organizations, and agencies constantly take decisions that influence community risk, and hence the probability that any safety goals will be met. Much of this policy design and implementation is carried out for economic, social, or aesthetic reasons which are not necessarily compatible with the goal of risk reduction. Every community is also influenced by a variety of regulatory agencies and by organizations such as hospitals, police, and fire departments, which are specifically concerned with reducing life and/or property losses.

In many instances what is needed is an organization, such as a safety committee, which will attempt to develop an overview of the risk implications of all of these activities. This is necessary so that an effort can be made to coordinate them and prevent unnecessary risk taking. In some cases this can be achieved by evaluating environmental impact statements and presenting position papers at inquiries and public hearings. In other instances, the safety plan coordinator must seek to mitigate danger by personal contact through suggestion and constructive criticism. In either case, the six basic principles of safety in design should be stressed wherever appropriate.

References

Adcock, H.W. 1978. The engineer's obligations as related to man-made hazards. *ASCE-ICE-CSCE 1978 Joint Conference on Predicting and Designing for Natural and Man-Made Hazards.* American Society of Civil Engineers, New York, pp. 39–48.

Ansell, P. R. 1979. Tagging explosives to counter crime. *Emergency Planning Digest,* **6(1)**:3–6.

Blume, J.A. 1978. Predicting natural hazards—State of the art. *ASCE-ICE-CSCE 1978 Joint Conference on Predicting and Designing for Natural and Man-Made Hazards.* American Society of Civil Engineers, New York, pp. 1–10.

Canadian Press. 1978. SOS system tested for cars, *Victoria Times,* March 10, p. 36.

Canadian Press. 1979a. Computer content with a handshake. *Victoria Times,* March 30, p. 10.

Canadian Press. 1979b. All except wash dishes. *Victoria Times,* June 9, p. 35.

Chandler, T., and G. Fox. 1974. *3000 Years of Urban Growth.* Academic Press, New York.

Council for Science and Society. 1977. *The Acceptability of Risks.* Barry Rose, London, 104 pp.

Davy, H. 1979. Electronic device notifies neighbour in emergency. *Victoria Times,* March 3, p. 12.

Dickson, P. 1976. *The Electronic Battlefield.* Indiana University Press, Bloomington, Indiana, 244 pp.

Fingas M. F., and C. W. Ross. 1977. *An Oil Spill Bibliography. March, 1975 to December, 1976.* Environmental Protection Service, Department of Fisheries and Environment Canada, 112 pp.

Fisher, B. E. 1975. Emergency helicopter rescue scheme for high-rise buildings. *Emergency Planning Digest,* **2(6)**:8–13.

Fisher, B. E. 1976. Underground disaster. *Emergency Planning Digest,* **3(1)**: 14–22.

Harmathy, T. Z. 1977. *Building Design and the Fire Hazard.* National Research Council of Canada, Division of Building Research, Technical Paper No. 805, pp. 127–144.

Hewitt, K., and I. Burton. 1971. *The Hazardousness of a Place: A Regional Ecology of Damaging Events.* University of Toronto Press, University of Toronto Department of Geography Research Series Publications, Toronto, Ontario, 154 pp.

Hughes, E. 1973. Pink was the color of death. *Reader's Digest,* December, pp. 184–190.

Hutcheon, N. B. 1974. *Trends in the Regulation of Safety and Performance in*

Buildings. National Research Council of Canada, Division of Building Research, Technical Paper No. 425, 9 pp.

Johnstone, A. R. 1978. Transportation of dangerous goods: Off to a good start. *Emergency Planning Digest*, **5(4)**:2–4.

Knoll, F. 1978. Discussion of Jack Rodin's predicting man-made hazards—State of the art. *ASCE-ICE-CSCE 1978 Joint Conference on Predicting and Designing for Natural and Man-Made Hazards*. American Society of Civil Engineers, New York, pp. 33–38.

Legget, R. F. 1973. *Cities and Geology*. McGraw-Hill, New York, 624 pp.

Miller, A. 1973. Vandalism and the architect. *In:* C. Ward (Ed.), *Vandalism*. Architectural Press, London, pp. 96–111.

Owens, E. H., A. Leslie, and Associates. 1977. *Coastal Environments, Oil Spills and Clean-up Programs in the Bay of Fundy*. Environmental Protection Service—Atlantic Region, Department of Fisheries and Environment Canada, 175 pp.

Pauls, J. L. 1975. *Fire Safety and Related Man-Environment Studies*. National Research Council of Canada, Division of Building Research, Paper No. 695, pp. 386–394.

Pauls, J. L. 1977. *Management and Movement of Building Occupants in Emergencies*. National Research Council of Canada, Division of Building Research, Paper No. 788, pp. 104–130.

Pugsley, A. 1973. The prediction of proneness to structural accidents. Cited in Rodin (1978).

Rodin, J. 1978. Predicting man-made hazards—State of the art. *ASCE-ICE-CSCE 1978 Joint Conference on Predicting and Designing for Natural and Man-Made Hazards*. American Society of Civil Engineers, New York, pp. 21–31.

SCICON. 1972. *Safety in Football Stadia: A Method of Assessment*. Scientific Control Systems Ltd., London.

Sullivan, W. 1979. Tangshan, 1976: Worst destruction since Hiroshima. Report from *New York Times*, *Victoria Times*, June 27, p. 5.

Tamura, G.T. 1954. Experimental Studies on Pressurized Escape Routes. *American Society of Heating, Refrigerating and Air-Conditioning Engineers Inc. (ASHRAE) Transactions* **88(2)**:224–237.

Texas Coastal and Marine Council. 1978. *Model Minimum Hurricane-Resistant Building Standards for the Texas Gulf Coast*. The Texas Coastal and Marine Council, Austin, Texas.

United States National Commission on Fire Prevention and Control. 1973. *America Burning*. U.S. Government Printing Office, Washington, D.C.

Ward, C. 1973. *Vandalism*. Architectural Press, London, 327 pp.

Working Group on Earthquake Hazards Reduction. 1978. *Earthquake Hazards Reduction: Issues for an Implementation Plan*. Office of Science and Technology Policy, Executive Office of the President, Washington, D.C., 231 pp.

5
Predicting and Preventing Disaster

With the refinement of our techniques for forecasting and planning, we are coming to realize that the image we hold of our future is itself an important element of that future. The expectations we arouse become a strong motivating force in realizing them.

Pierre E. Trudeau (1970)

While limited to only one past and one present, every society faces a multiplicity of potential futures (Sewell and Foster, 1976). Unfortunately for some communities, disaster looms large in many such visions. Every safety plan coordinator and his associated committee must have an image of the future, therefore, which is based on a clear understanding of the potential catastrophes that their society faces. Without it, optimum warning systems, adequate disaster plans, and viable reconstruction guides cannot be developed. Similarly, such a vision will also permit steps to be taken to reduce risk and so mitigate the potential for death and destruction.

The fact that the future does not yet exist has not discouraged the search for methods of knowing and controlling it. These range from the supernatural, divined by Tarot cards and horoscopes, to the scientific, involving such techniques as the application of computer simulations and the Delphi questionnaire approach.

Attempting to predict and respond to potential disasters is essentially a branch of futurology (Henchey, 1978). Such research is generally taken to involve "the projection of present trends into the future, predictions of future events or of the state of society at some future date, and long range planning for organizations, institutions or societies" (Coates, 1972). There are a wide range of techniques currently being employed by those involved in futurology (Table 5.1). It is probably true to say that all of the 27 methodologies listed in this table could be applied to assist a safety plan coordinator in developing an image of the potential future disasters that face his or her community. Nevertheless, some of these techniques have rarely, if ever, been used for this purpose, while others have been

Table 5.1. Techniques Currently Employed in Futurology

1. Scenario building	15. Operations research
2. Delphi technique	16. Survey research
3. Simulation/gaming	17. Decision matrices
4. Trend extrapolation	18. Growth curves
5. Scale modeling	19. Interviewing
6. Analogue modeling	20. Operational gaming
7. Computer simulation	21. PERT adaptation
8. Cross-impact analysis	22. Role-play gaming
9. Correlation plotting	23. Speculation (disciplined)
10. Expert position papers	24. Values analysis
11. Relevance trees	25. Q analysis
12. Analogy	26. Morphological analysis
13. Economic projection	27. Lateral thinking
14. Objectives trees	

Source: Modified after Coates (1972).

shown to have enormous value in disaster research. It is with these latter approaches, simulation, scenario building, role-play gaming, the Delphi technique, and field exercises that the remainder of this chapter is concerned.

Disaster Simulation Models

Simulation models are an important method of investigating the development of the potential for disaster through time. These are normally of three types: scale, analogue, and mathematical (Chorley and Kennedy, 1971).

Scale Models

Scale models are used most frequently by engineers, hydrologists, and geomorphologists to study the evolution of physical systems, such as rivers or sea coasts. They utilize equivalent real world materials which have been scaled down in size both geometrically and kinematically. Fine sand or clay may be used in a stream-table or flume, for example, to simulate soil erosion and sedimentation. Kaolin can be mixed to a consistency that allows the modeling of glacial surges. Disasters may be simulated by simply introducing scale models of buildings and other aspects of land use into such physical models. The impact of the disaster agent can then be photographed, measured, and recorded.

Scale models constructed in the laboratory have been used to study both past disasters and the future potential for destruction. On 21 October 1966 a slide occurred in colliery waste at Aberfan in South Wales. This overran a school killing 116 children and 5 teachers; in all, 144 people lost their lives. Field observations, laboratory analysis, and the use of a hardware model, constructed at a scale of 1:100, allowed the identification of the key variables responsible for this slide. These were found to be the height of the colliery waste, high water pressure due to heavy rainfall, and the presence of a shear surface within the tip, which had been created by earlier movements (Bishop, Hutchinson, Penman and Evans, 1969).

Scale models may also be useful in helping to prevent future disasters. Port Alice, on the northwest coast of Vancouver Island, British Columbia, is a company town that has been subjected to severe mudflows and, indeed, appears to be built on a debris fan. Damage totaling some $800,000 was caused on 15 December 1973 by a mudflow originating at the 2300 ft level on steep slopes above the town. As a result of this mass movement, one $39,000 home was demolished, nine others were rendered uninhabitable, 20 vehicles were damaged, and the town's storm drains and gas lines became inoperable. A further slide occurred in 1975 (Scanlon, Jefferson and Sproat, 1977).

Theoretical studies have been conducted since and a scale model of the town and adjacent slopes constructed. Mudflows were reproduced using bentonite mud with a similar viscosity to the coarse, bouldery gravel actually involved. On the basis of the information obtained, a diking system was constructed at a cost of $250,000. This was designed to protect the town from slides up to 2.5 times the volume of those recently occurring. Future mudflows will be diverted into unsettled areas (H. W. Naismith, 1978, personal communication).

The possibility of large landslide-generated waves overtopping the Libby dam, Montana, was investigated using a hydraulic model. This was built at an undistorted scale of 1:120, at the U.S. Army Corps of Engineers Waterways Experiment Station, Vicksburg, Mississippi. The effects of landslides into Lake Koocanusa from 160 to 2100 ft upstream of the dam were reproduced using the equivalent of volumes varying from 900,000 to 4,750,000 cubic yards (Davidson and Whalin, 1974).

Many other hydraulic models have been built to assist in predicting disaster potentials. These include the model of the Hudson River, constructed at Vicksburg to study sedimentation in the New York City area. Using this structure to simulate the processes involved, numerous solutions to this problem were suggested. They included the construction of sediment basins to trap the shoal materials, closure of the Harlem River to prevent the mixing of salt water with that from the Hudson, and the deepening and widening of specific sections of the channel (Panuzio, 1968). The largest working hydraulic simulation of the U.S. Army Corps of Engineers, the San Francisco Bay and delta model, is located at Sausalito, California. This covers $1\frac{1}{2}$ acres and allows the study of threats to shipping, salt water intrusion into farmland, and other related hazards (Gibson, 1976). Chapon (1961) described a hydraulic model of part of the Seine estuary, constructed at a horizontal scale of 1:800 and a vertical scale of 1:100. This was built to assist in predicting the impact of engineering works and the evolution of channel depths over a period of decades. The latter was of major significance to the economic life of the port of Rouen, France. Similar models have also been built of the Severn estuary in Britain (Allen, 1947).

Other hazards which have been simulated using scale models include earthquakes and tsunamis. A large accelerator, for example, has been built at the Tsukuba Research Center in Education City, Japan. This model is capable of reproducing seismic waves, similar to those usually experienced during earthquakes. Structures up to 200 tons in weight can therefore be subjected to a vertical acceleration of 1.0 *g*, allowing prediction of the impact of such forces on the variety of building types (Japanese National Research Center for Disaster Prevention, 1971).

Whalin, Bucci, and Strange (1970) have described a scale representation of the harbor at San Diego, California. This was built to investigate the impact of wave periods ranging from 40 to 186 sec and deep-water wave heights of from 12 to 45 ft. Such waves would be associated with impulsive sources, such as localized seismic disturbances, an explosion, a massive landslide, or the impact of a meteorite. They concluded from use of the model that waves of this nature would cause extensive inundation of the Silver Strand, the city of Coronado, and part of North Island. It was thought unlikely that any craft would survive them in the surf zone.

Models are also being developed of certain meteorological hazards. Perhaps the most significant, from the viewpoint of this volume, are those of tornadoes. These twisting columns of air can now be modeled in the laboratory. Experimental vortices have been produced in liquids and gases and in liquids bubbled with gases to simulate buoyancy forces. There are some difficulties in the laboratory involving reproducing the effects of gravity, latent heat release, and absence of sidewalls. Nevertheless, both multiple and single vortices have been created in one apparatus by the control of air inflow angle and depth of the inflow layer. The vortex family is produced when the circulation of inflowing air is sufficiently increased by faster rotation of an encircling screen (Kessler, 1977).

It is estimated that in the United States alone, an average of 3 billion tons of soil are eroded each year, containing 50 million tons of plant nutrients. If fertilizers were to be used to replace these the cost would be some $6.8 billion. In addition, enormous expenses are also incurred as the result of sedimentation, over half a billion cubic yards being dredged annually to maintain navigable waterways (Beasley, 1972). Numerous models have been built to simulate the rates of erosion associated with precipitation of differing intensities. Such rainfall simulators, for example, have been used to study the relative erodibility of loess-derived soils in southwestern Iowa (Schmidt, Schrader and Moldenhauer, 1964). The value of such equipment has been discussed by Bryan (1967).

In summary, scale models may be of considerable use to safety plan

PLATE 9. Dust storm, Mantes village, near Liberal, Kansas. It is estimated that, in the United States alone, an average of 3 billion tons of soil are eroded each year (American Red Cross photograph).

coordinators and their committees as aids in determining why past disasters have occurred. If they are successfully constructed they should be capable of reproducing many of the characteristics of such former events. Once successfully tested in this manner, valid models can then be used to assist in predicting the future. With the aid of a photographic record they can then be used in teaching and training and as aids in developing warning systems and disaster plans.

Analogue Models

Analogue models use materials which have some very different properties from those being simulated, but have certain characteristics in common with them. Electrical groundwater analogues, for example, use electric potential to represent the height of the water table and current to simulate the flow of groundwater. Such a system may be used to model the impact of drought. Electrical analogue models have certain advantages over scale models. They are usually cheaper, easier, and quicker to construct and operate. An electrical analogue gives almost instantaneous solutions to complex problems. These can be displayed on a cathode ray tube allowing the parameters of the model to be rapidly adjusted. This permits a closer fit to be achieved between the behavior of the real world system and the analogue (Chorley and Hagget, 1967). Einstein and Harder (1961) investigated the tidal flows of the delta region of California using an electrical analogue model, while Harder (1963) extended the concept to study the Kansas River basin. In this area he simulated flood control systems with the objective of determining the optimum phasing of reservoir sluice operation during a flood. Discharges in tributaries and the main channels were represented by electrical current flow. This could be manipulated to represent a wide variety of rainfall conditions and resulting runoff. Because the speed of operation of this electrical analogue model was so rapid, only 0.03 sec, it was possible to evaluate a large number of alternative plans for emergency reservoir operation extremely quickly. More permanent solutions to repeated flooding problems could also be compared with relative ease.

Analogue models need not be electrical. A wide variety of these surrogates for disaster agents have been used. These include heat conductance to simulate stream flow (Appleby, 1956) and carbonated water to produce a model of a tornado vortex (Turner and Lilly, 1963). Blotting paper analogues have also been widely applied in studies of diffusion and growth and may have some value in modeling the spread of infection (Chorley and Hagget, 1967).

In general, however, analogue modeling has many intrinsic disadvantages and is rarely used for such practical purposes. Much more emphasis is now being placed on the development of computer based simulation models which counterfeit some aspects of the real world and allow predictions to be made about its future behavior (Chorley and Kennedy, 1971).

Computer Simulations

Computer simulations which permit relatively accurate predictions of potential disaster losses are invaluable management tools. Regardless of the hazard involved, the construction of such models involves four common steps. The first of these is an initial analysis of the physical characteristics of the hazard. This permits the subsequent development of a mathematical model capable of forecasting the severity and frequency of its impact. The approach taken is to develop a model that produces a spatial pattern of intensities with properly spaced contours which are consistent with the size, shape, and configuration of observed patterns (White and Haas, 1975). This distribution will be controlled by the magnitude of the event, modified by the impact of certain local variables. In the case of the earthquake hazard, for example, depth to bedrock, height of water table, and sediment type will cause variations in seismic intensity. Similarly, the spread of forest fires will be influenced by such factors as the availability of fuel, breaks in tree cover, location of water bodies, and steepness of slope angles. It is the distribution of these locally distinct variables that must be established when risk mapping is being carried out. The production of a single hazard–multiple purpose microzonation, which distinguishes spatial differences in the intensity of impact of a disaster agent, is then the first essential step in such a computer simulation.

To predict the damage and casualties such an uneven distribution of intensities will cause, it is necessary to know the geographical location and characteristics of the population and the value of the infrastructure that is at risk. This can be determined by surveys of building types, occupancy rates, and internal fittings. The distribution of communications, transportation, and utility networks and of crops and other land uses must also be established. Such information is used to produce a geographical representation of the society threatened by the hazard.

In the United States, existing general studies of natural hazards accept input data using a system of grid areas of about 35 square miles each. Approximately 85,000 of these are needed to represent the total land area of the 48 contiguous states. All these grid areas are addressed on a computer disk so that such property characteristics as number, value, type, exposure, and vulnerability can be stored for each (White and Haas, 1975). This system uses two measures of population-at-risk, the population size and the number and value of one-family dwellings. The Travelers Insurance Company is the pioneer in this field; using information from the 1970 Census its employees are allocating 203 million individuals and 47 million one-family dwellings to 85,000 such grid areas (Friedman, 1973; White and Haas, 1975).

Once these two steps have been taken, the models of the disaster agent and of the infrastructure and its inhabitants must be linked by a marix representing the loss relationship between property type and the intensity of impact. For example, it is known from widespread experience how much damage various types of buildings sustain from flood waters of differing depths and velocities, seismic shocks, or explosions of particular force (Rinehart, Algermissen, and Gibbons, 1976). It is therefore possible to predict the degree of destruction related to particular intensities of impact on specific buildings. In addition, for most hazards there is also a relationship between the length of the warning period, the magnitude of the destruction, and the associated casualty rate. Deaths and injuries can therefore be predicted for events of differing scale. Once such a matrix has been placed into memory, computer simulation can be used to apply the mathematical representation of the hazard to the geographical distribution of people and infrastructure. This produces a synthetic disaster experience which can be presented in terms of economic loss, degree of damage to particular buildings, fatalities and injuries sustained, or stress experienced. Such information is normally presented in map form and also in tabulations so that both the spatial distribution and gross impact are immediately obvious. Computer simulations can be undertaken using the same four stages for any hazard. In a fifth step, their validity can also be checked by using them to simulate losses from past catastrophes. The more valid the simulation, the closer are the actual and predicted consequences.

One early attempt at such a simulation process occurred in the United States where the National Oceanic and Atmospheric Administration was commissioned by the Office of Emergency Preparedness to undertake an assessment of potential seismic damage in the San Francisco Bay area. This study examined the consequences of earthquakes of Richter magnitude 6.0, 7.0, and 8.3 occurring along the San Andreas and Hayward faults. For each event an isoseismal map was prepared to show differences in associated intensities (Figure 5.1). Important aspects of the infrastructure such as hospital facilities, ambulance service groups, and utility networks were also mapped and scored in terms of their earthquake resistance. The interaction of the released physical forces and these structures was then analyzed, and the damage was estimated (U.S. Office of Emergency Preparedness, 1972).

One of the most detailed computer simulations carried out to date was

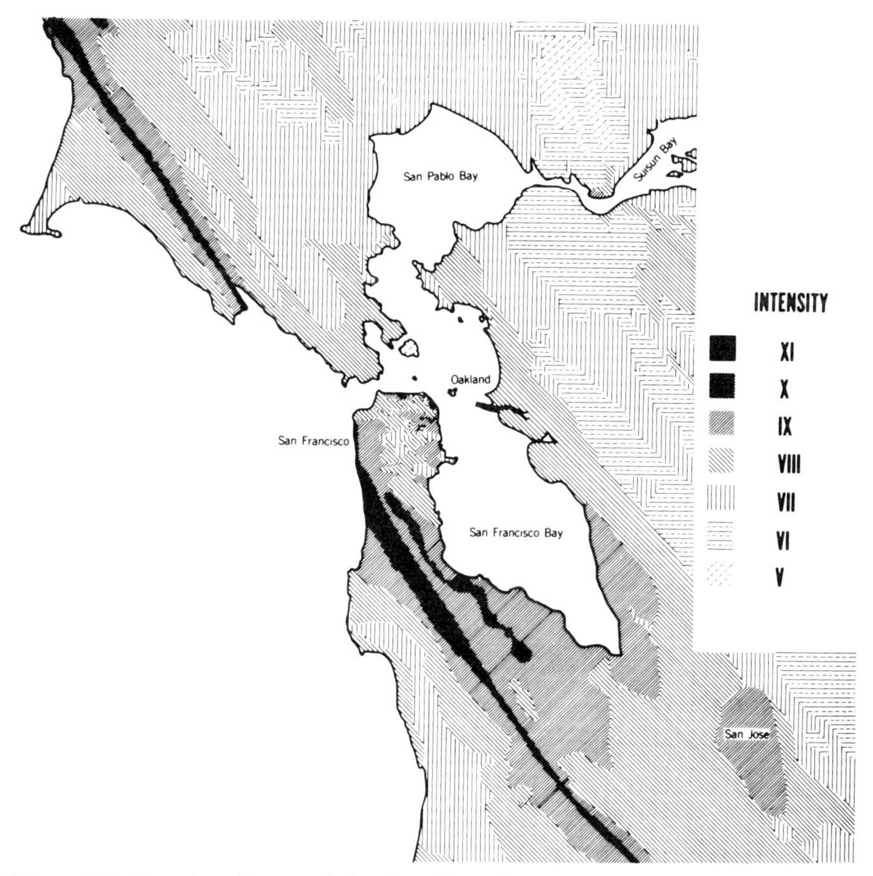

Figure 5.1. Isoseismal map of the San Francisco Bay Area showing estimated modified Mercalli intensities for a major (8.3) earthquake on the San Andreas fault (after Office of Emergency Preparedness, 1972, a predecessor of the U.S. Federal Emergency Management Agency, *Disaster Preparedness*).

that undertaken for Victoria, British Columbia, by Foster and Carey (1976). This attempted to predict seismic damage on virtually a building by building scale. It is discussed in detail here so that its methodology can be fully appreciated for application elsewhere.

The first step in this simulation involved establishing the average seismic intensity likely to be associated with the 100-year earthquake and the prediction of variations from it. The nature of earthquake risk in Canada has been described in detail by Whitham, Milne, and Smith who, in 1970, published a seismic zoning map, based on an analysis of all earthquakes affecting the country since 1899. The risk at any one location was derived from all known earthquakes which had influenced that site, and was calculated by determining the maximum peak horizontal ground acceleration of each seismic event. In this manner, Canada was subdivided into four zones, those areas where seismic risk was greatest being placed in Zone 3. Since Victoria is situated in this zone an acceleration of 6% gravity or greater, having an annual probability of exceedance of 1 in 100 should be anticipated. This approximates to at least a modified Mercalli intensity of VII on "normal" ground, with intensities reaching as high as VIII in more vulnerable areas where bedrock is deeply buried, once a century.

This estimate contrasts with the earlier work of Milne and Davenport (1969) who carried out an analysis of 1479 earthquakes influencing western Canada between 1899 and 1960. Their study postulated a peak horizontal ground acceleration of 10.7% gravity, for a hundred year return period; that is, approximately intensity VIII on "normal" ground, IX on unstable sediments, and VII on bedrock.

Wuorinen (1976) discussed those physical factors which influence the distribution of earthquake intensities at the local level. His microzonation of Victoria was used as input data for the simulation discussed here (Figure 3.1). A few further qualifying comments appear necessary. Wuorinen's microzonation is limited to anticipated intensities associated with primary shock. Damage from aftershocks and other hazards associated with earthquakes, such as sand boils and mass movement, have not been included. Similar limitations, therefore, also apply to Foster and Carey's (1976) simulation.

Once the physical characteristics of the hazard had been determined it was necessary to quantify the infrastructure at risk, the second step in the simulation process. Twenty-four significant categories of land use and building type were established by a detailed survey of the literature of past seismic disasters. These classes were based on response differences, noted during seismic ground motion experienced elsewhere. The classes identified are illustrated in Table 5.2. Since no up-to-date information was available on land use in Victoria, a building-by-building survey was undertaken. To facilitate this, a grid system was used. Each rectangle so produced represented an area of approximately 40 by 58 meters, within

Table 5.2. Mean Damage Ratio Matrix

Structural classification	Modified Mercalli intensity					
	VI	VII	VIII	IX	X	XI
No construction; parks, cemeteries	0.0	0.0	0.0	0.0	0.0	0.0
Asphalt; playgrounds, tennis courts, parking lots	1.0	15.0	40.0	66.0	100.0	100.0
One to three story residential; wood frame; less than 6000 sq ft	0.3	1.25	8.25	12.0	20.0	50.0
One to three story residential; masonry, less than 6000 sq ft	2.0	4.0	20.0	70.0	98.0	100.0
Single story business and personal service occupancy; wood frame; less than 6000 sq ft	0.5	1.5	9.0	15.0	25.0	60.0
Single story business and personal service occupancy; masonry and hollow brick, less than 6000 sq ft	1.0	3.0	20.0	75.0	99.0	100.0
Two to three story business and personal service occupancy; wood frame; less than 6000 sq ft	1.0	2.5	12.0	22.0	35.0	75.0
Two to three story business and personal service occupancy; masonry and hollow brick; less than 6000 sq ft	2.0	5.0	25.0	85.0	100.0	100.0

Table 5.2. *(continued)*

Structural classification	Modified Mercalli intensity					
	VI	VII	VIII	IX	X	XI
Medium and low hazard industrial buildings; three storeys or less; wood frame	2.0	4.0	20.0	92.0	100.0	100.0
Medium and low hazard industrial buildings; three stories or less; masonry or hollow brick	2.25	4.5	22.5	99.0	100.0	100.0
Medium and low hazard industrial buildings; three stories or less; steel frame	0.75	1.8	8.5	45.0	85.0	100.0
Buildings with a ductile moment resisting space frame	4.0	8.5	18.0	45.0	65.0	85.0
Buildings with a dual structural system consisting of a ductile moment resisting space frame and ductile flexural walls	4.5	9.5	20.0	50.0	72.0	94.0
Buildings with a dual structural system consisting of a ductile moment resisting space frame and sheer walls; also buildings with ductile flexural walls	5.0	10.5	22.0	54.0	80.0	100.0
Buildings with reinforced concrete sheer walls	0.13	1.4	10.0	45.0	95.0	100.0

Buildings with unreinforced masonry and unreinforced concrete frames and walls	2.5	8.0	25.0	98.0	100.0	100.0
Subaerial bulk fuel storage tanks	0.0	9.0	40.0	60.0	90.0	100.0
Storage tanks and contents other than fuel tanks; elevated tanks	0.0	12.0	60.0	99.0	100.0	100.0
Smokestacks, standpipes, and similar structures not supported by a building	20.0	60.0	100.0	100.0	100.0	100.0
Dams; reservoirs	0.0	2.0	30.0	70.0	80.0	96.0
Poured concrete walkways, piers, and retaining walls	0.0	2.0	25.0	65.0	75.0	90.0
Steel frame (through truss) bridges	0.0	3.0	35.0	75.0	100.0	100.0
Reinforced concrete bridges	0.0	8.0	40.0	80.0	100.0	100.0
Wood pile piers; wood pile wharves; wood pile bridges	0.0	0.0	10.0	25.0	35.0	66.0

Source: Foster and Carey (1976).

which one predominant land use was identified. Because of the large scale involved this was commonly a single building. Each structure or land use was assigned to the appropriate category and is represented as such in Figure 5.2.

The third step in developing a simulation of potential earthquake destruction in Victoria involved producing a potential damage matrix. While the accurate prediction of damage to individual buildings is extremely difficult to make, trends in response to seismic events by particular building types have been identified elsewhere (Whitman, 1973; Steinbrugge, 1970). These can be used to anticipate general patterns of damage associated with earthquakes of differing magnitudes.

The expense of repairing a building, expressed as a percentage of its cost of total replacement, is known as the damage ratio. A mean of these ratios for all buildings within a particular category is known as the mean damage ratio (MDR) and was developed for all 24 structural types used in the land use classification. This mean value replaces a full set of damage probabilities with a single "average" figure. Table 5.2 illustrates the resulting matrix. It consists of a mean damage ratio for each structural classification, for earthquakes generating modified Mercalli intensities of VI to XI on "normal" ground. Values one intensity above and one below the range actually simulated for Victoria were necessary because of the amplification of ground movement in areas of deep sediments and its reduction in areas where bedrock is at or near the surface.

The information used to construct the mean damage ratio matrix, illustrated in Table 5.2 was derived from a variety of independent sources. For some structures, mean damage ratios have already been published, detailed damage statistics which allow their calculation have been made available for others by Steinbrugge (1970). In a few cases, only general descriptions of damage, caused by particular earthquakes to certain structures, were available. In this case, estimates of mean damage ratios were made using apparent trends in reactions of similar buildings for which more complete information had been published.

The damage that various structures can be expected to experience during an earthquake varies considerably, depending upon design strategy, constructional material, location, and age. The quality of construction within a particular engineering system can also be a major factor. Unreinforced masonry and concrete structures generally suffer greater damage during earthquakes than wood-frame structures, primarily because of their greater mass. Wood-frame dwellings tend to be more flexible when subjected to vibration and have a high survival rate, even after very large seismic shocks. Failure of wood-frame structures under four stories during an earthquake is most often due to a lack of lateral force bracing at foundation level, or poor condition of existing bracing, with a resistant movement of the entire structure off its foundations. Buildings using

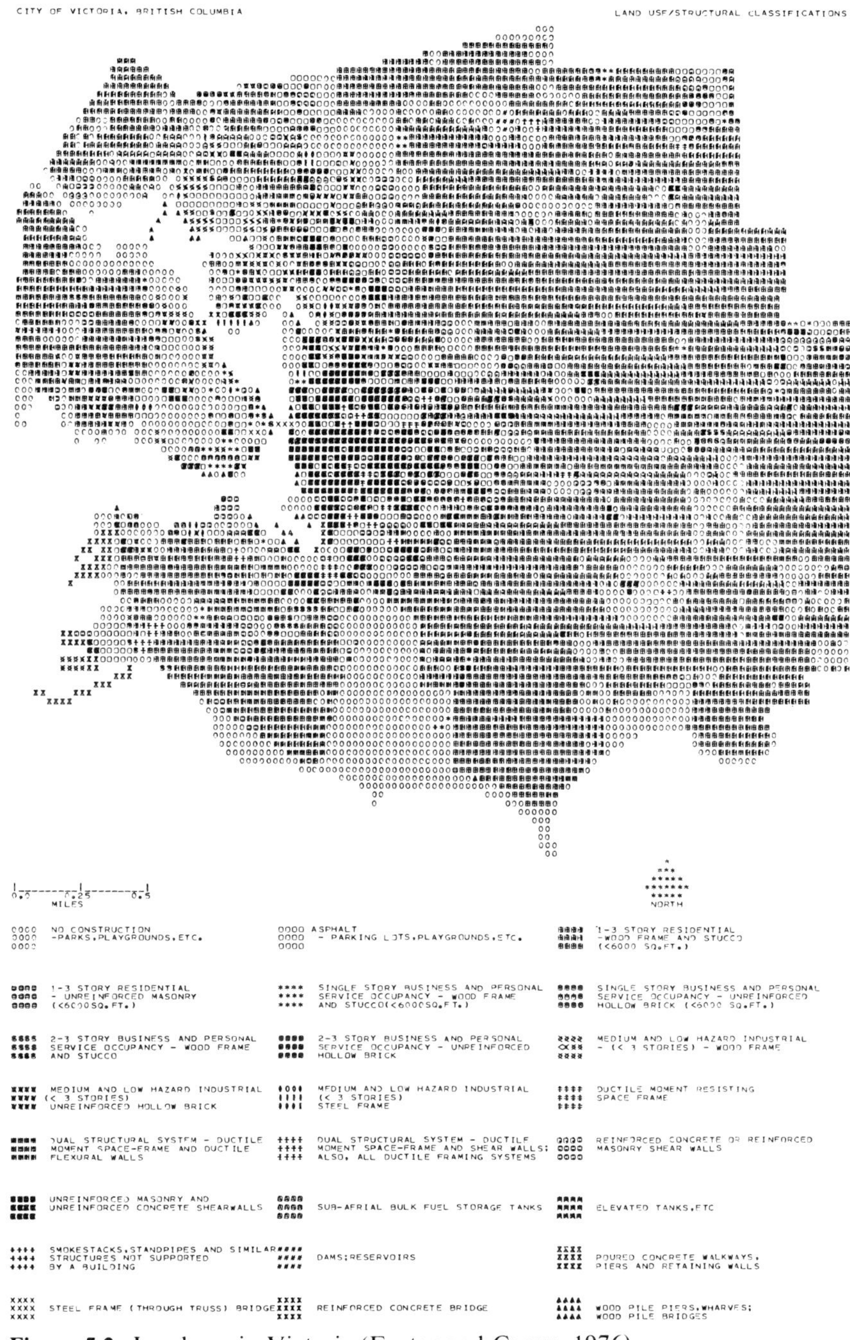

Figure 5.2. Land use in Victoria (Foster and Carey, 1976).

modern earthquake-resistant design strategies, and structures such as bridges or fuel tanks, tend to have mean damage ratios between these two extremes. Within the damage matrix, areas classified as "no construction" are shown to experience, of course, no structural damage. However, surface upheaval, mass movement, and other related phenomena can be expected at intensities of VIII and above.

This matrix (Table 5.2), showing the "average" anticipated damage for particular earthquake intensities, together with the land use survey and the microzonation map allowing prediction of spatial distribution in intensities permitted the simulation of earthquake damage for the City of Victoria. Examples are presented as Figures 5.3 and 5.4. The former illustrates expected damage during the 100-year earthquake. Over much of the city, little or no severe structural damage is anticipated. Many exceptions, however, do appear to be probable. For example, heavy damage (between 20–65% of replacement value) can be expected to occur in close proximity to the Inner Harbour. The potential for moderate damage (7.5–20% of replacement value) is far more widespread. It can, for example, be expected to occur throughout much of the lower-lying parts of the Fairfield district and north of Ross Bay in the south of the city and west of Douglas and north of Finlayson Streets in the other extremity. Numerous potential moderate damage pockets can also be identified including the area west of City Hall. Many of the one to three story woodframe and stucco residential buildings in the city would, however, suffer no appreciable damage in an earthquake generating such intensities.

Figure 5.4 illustrates the impact of an earthquake capable of intensity VIII on "normal" ground. Naturally, anticipated associated damage would be greater. Isolated examples of collapse (99–100% of replacement value) could be expected. Similarly much of the downtown area would suffer damage ranging from moderate to sufficient to make repair impossible or uneconomic. Other pockets of heavy damage might be expected in many locations, particularly at the western end of Johnson Street and south of Rock Bay. Although some buildings could be expected to suffer light or no damage, the city as a whole would be extremely hard hit by an earthquake capable of generating intensities of this size.

If a survey had been undertaken to establish building occupancy rates at particular times of day, it would also have been possible to predict deaths and injuries from such seismic events. This could have been achieved by the application of the information given in Table 5.3, which relates the damage state of a building to the casualties that this causes. For example, if a collapsed building had an occupancy of 100, 20 of these could be expected to die and the remainder would be injured. By applying the formula given in Chapter 2, this information could be converted, if required, to community stress units. Estimates of occupancy rates were used to obtain such a stress total and are presented in Chapter 6.

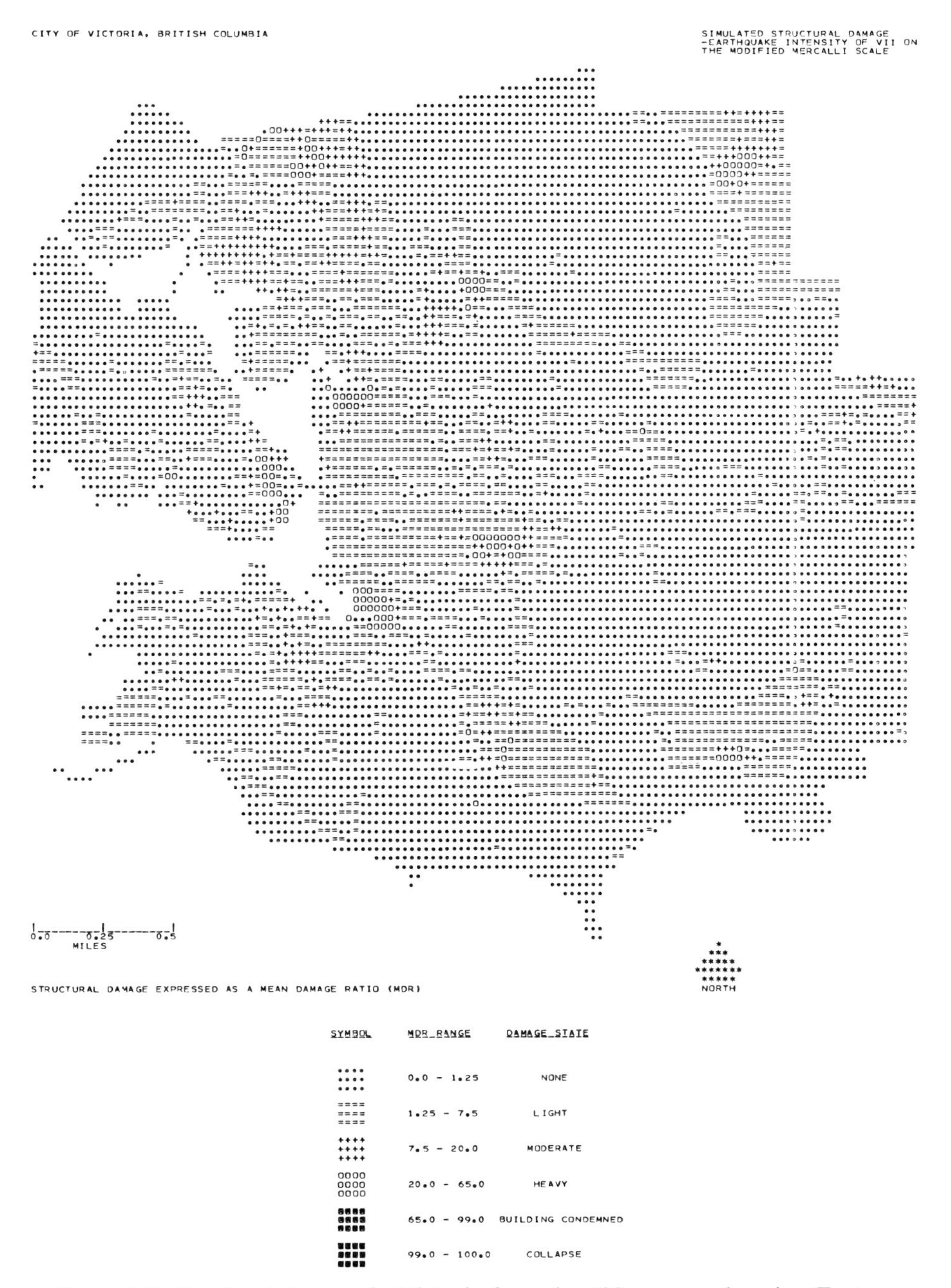

Figure 5.3. Simulated damage in Victoria from the 100-year earthquake (Foster and Carey, 1976).

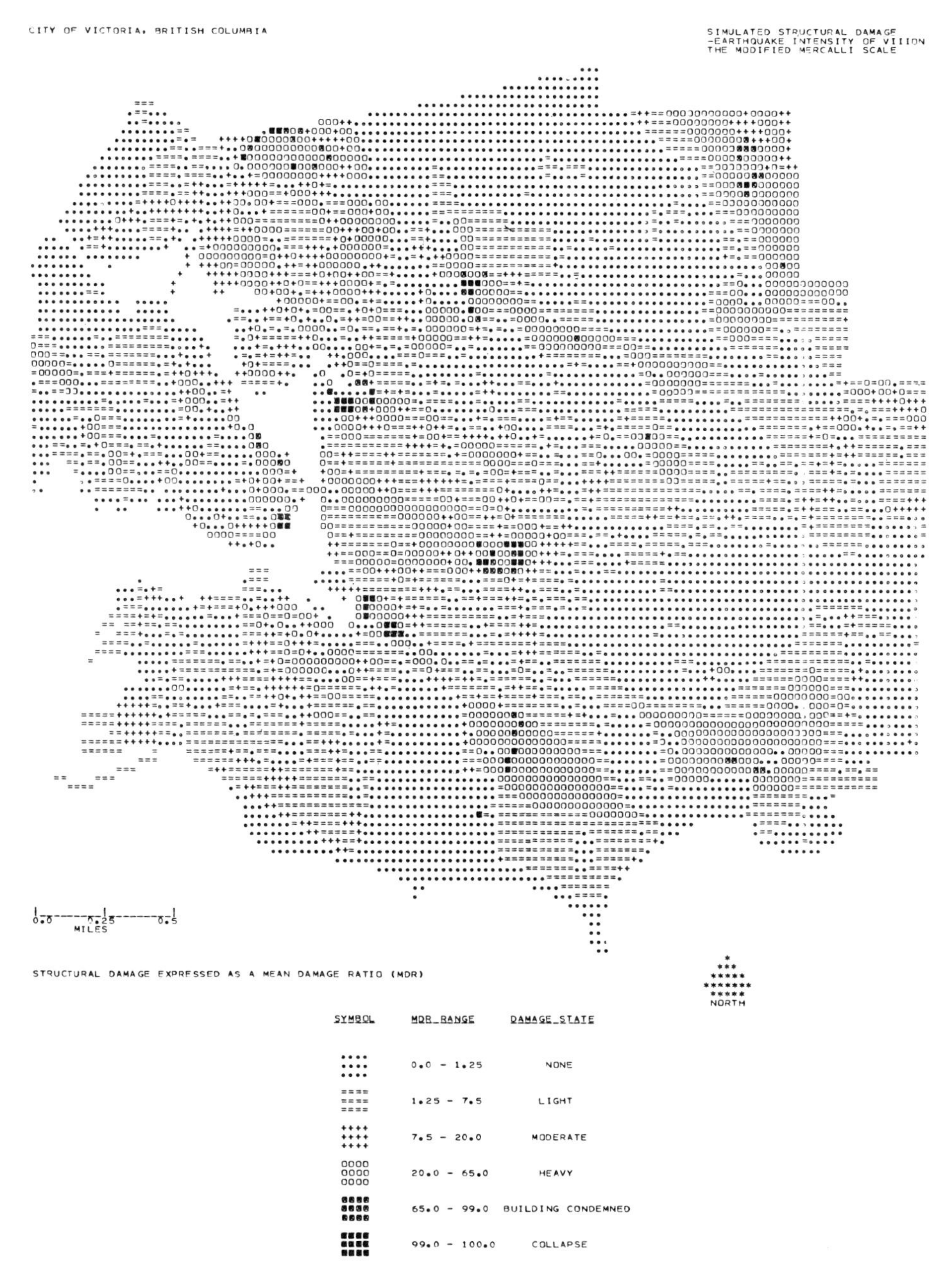

Figure 5.4. Simulated damage in Victoria from an earthquake reaching a modified Mercalli intensity VIII on normal ground (Foster and Carey, 1976).

Table 5.3. The Relationship Between Mean Damage Ratios and Casualties

Damaged state of building	Fraction dead	Fraction injured
None	0	0
Light	0	0
Moderate	0	1/100
Heavy	1/400	1/50
Building condemned	1/100	1/10
Collapse	1/5	4/5

Source: Whitman *et al.* (1973).

A final step in this process of simulation is prediction of the past, so that forecasted losses can be checked against actual records. This has not been accomplished in Victoria but Friedman (1969) has attempted to compare predicted seismic damage in California by contrasting actual and simulated intensities for both the 1906 San Francisco and the 1956 Saint Helena earthquakes (Figures 5.5 and 5.6). Similarly, McIntyre (1979) has recently produced a computer simulation of large storm surges that occur along the west coast of the British Isles. These are of particular importance since they affect Milford Haven with its major oil tanker facilities. Predictions produced by this model were compared with actual surges experienced there in January 1976. The results showed excellent correlation, demonstrating that useful first-order accuracies in prediction can be achieved by mathematical simulation.

The enormous data bank developed by the Travelers Insurance Company is being used to simulate the potential impact of several major hazards in the United States. To determine the loss potential from hurricanes from Texas to Maine, for example, a century of available data on their characteristics, such as storm paths, speed, size, and intensity, has been computerized. This has been used to develop geographical patterns of maximum wind speeds for each past hurricane. These are then aggregated for 25-year periods and their effects on the current geographical distribution of properties estimated for insurance purposes. The same company has been active in developing a mathematical model to simulate losses from tornadoes, hailstorms, and thunderstorm-wind hazards (White and Haas, 1975).

Computer simulation was also used by the U.S. Department of Housing and Urban Development to estimate loss potential for coastal flooding during development of the Joint Insurance Industry/Federal Government National Flood Insurance Program (Kaplan, 1971-1972). This model is being refined to produce a more realistic envelope of maximum surge

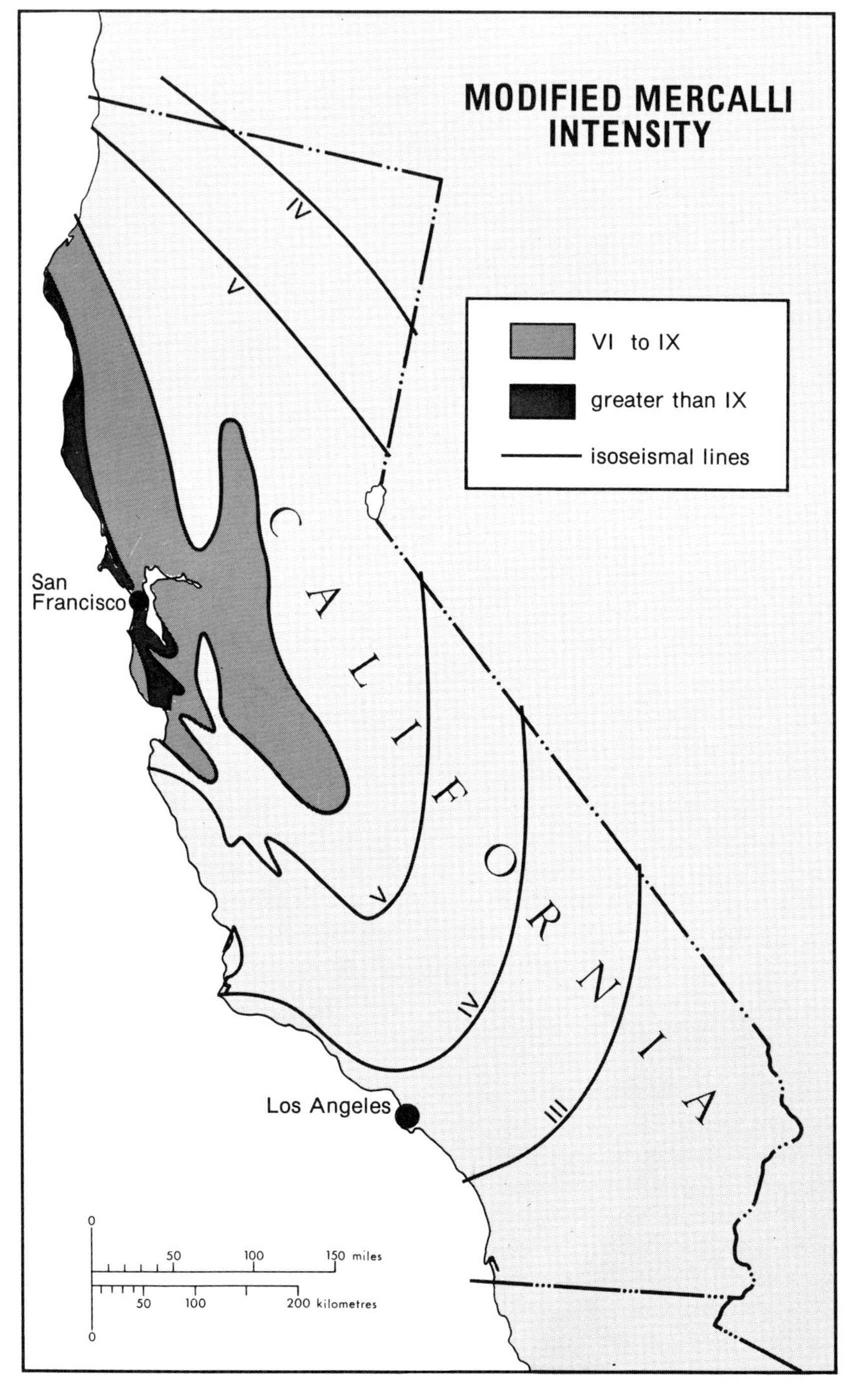

Figure 5.5. Actual isoseismal pattern of the 1906 San Francisco earthquake (Friedman, 1969).

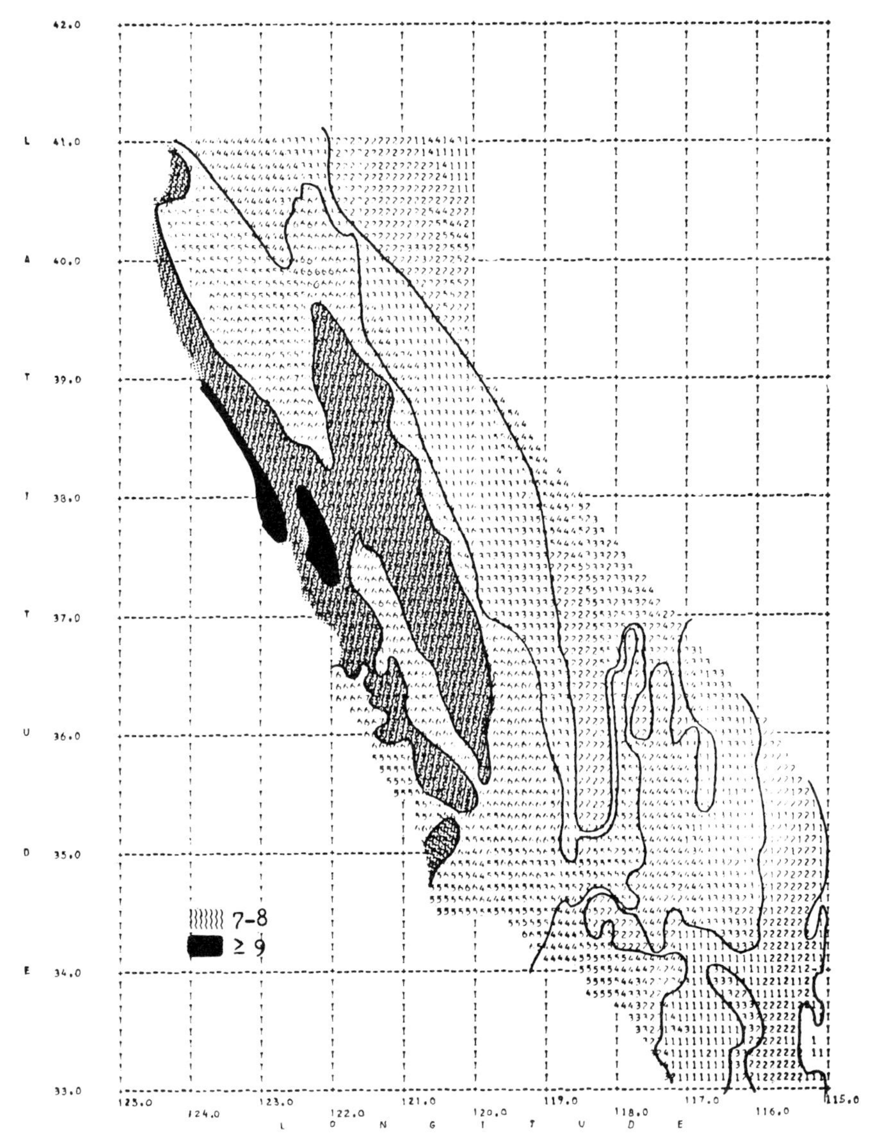

Figure 5.6. Simulated isoseismal pattern of the 1906 San Francisco earthquake (Friedman, 1969).

depths associated with hurricane passage, based on the wind profile developed from a hurricane model and local bathymetry conditions. For the same insurance study, inland flooding was also simulated by a computer model, designed for the Department of Housing and Urban Development, to establish the potential losses this could cause to urban areas (Friedman and Roy, 1966).

It must be stressed that although this computer modeling approach has been largely limited to simulating the impact of natural hazards, it could easily be adapted to forecast the potential damage and casualties associated with the impact of man-made agents. Although the input data would naturally vary with hazard and location, the basic steps would remain unaltered.

Simulations of this type have numerous practical uses. No disaster plan can be developed realistically without them since they provide an overview of the situations with which the plan seeks to deal. Such simulations, for example, can be used to predict which buildings are likely to be destroyed or seriously damaged and by what magnitude of disaster. They can also be used to forecast which roads, railways, telecommunications centers, and airports will probably be closed by damage, and how and where individuals will be killed or injured. The serviceability of hospitals, fire, and police stations and other public buildings can also be predicted. Obviously it is against the background of such disruptions that disaster plans must be formulated. One should not prepare to deal with 50 casualties if 500 is a more realistic estimate nor can vehicles and buildings that are likely to be destroyed on impact be counted upon to assist in dealing with the aftermath. Such simulations are also of value in predicting the occurrence of secondary hazards. Earthquake simulations, for example, permit forecasts to be made of where gas lines are most likely to fracture. Flood simulations can pinpoint when and where sewage plants and refuse disposal sites are likely to lead to dangerous pollution. Those used to predict the spread of fire are valuable in allowing estimates to be made of when chemical storage facilities are likely to be threatened.

Computer simulations of disaster are also valuable in the design of warning systems since they permit a municipality or region to be subdivided into danger zones on the basis of expected losses. In this way, particular emphasis can be placed on warning and evacuation in those areas where losses are likely to be the greatest. Such simulations also allow predictions of the impact of potential disasters on the region's economy. To illustrate, recent simulations demonstrate that if the 1906 San Francisco earthquake had recurred in 1974 it would have cost the Bay area in excess of $13 billion, approximately one-half of which would have resulted from lost income due to regional economic recession (Cochrane, Haas, Bowden, and Kates, 1974).

The value of being able to predict accurately those regions which are at risk from particular hazards was clearly illustrated in the United States in 1969. During the winter of 1968–1969 the National Weather Service expressed concern over the depth of the snow pack determined by ground observations and satellite sensing. They predicted disastrous spring floods over much of the northern Midwest. At the direction of the President, Operation Foresight was launched to coordinate the efforts of federal, state, and local authorities to reduce losses and casualties.

On 13 March 1969 the NWS River Forecast Center at Kansas City, Missouri, issued preliminary crest forecasts for 225 points in the threatened area, and these were continuously updated. They did, however, permit the identification of those communities likely to suffer major damage. In these areas a wide variety of measures were implemented to reduce losses. The U.S. Army Corps of Engineers helped state and local authorities to get protective measures in place before flooding began. Technical assistance was provided. Contracts were awarded for levee construction, and sandbags, polyethylene sheets, pumps, and lumber were supplied to local work teams. Crop-dusting planes blackened lake and river ice to encourage premature melting. The U.S. Coast Guard moved boats and helicopters into the area to aid in evacuation. The Department of Agriculture advised farmers on how to minimize crop, livestock, and machinery losses. The Interstate Commerce Commission gave priority to the transportation of grain from threatened flood plains. States declared emergencies and rescheduled essential public services so they would continue during flooding. State Highway Commissions provided trucks to move personal property out of the area, while local authorities raised labor to construct defences. Numerous other agencies and charities marshalled their resources to prepare for the anticipated floods (U.S. Office of Emergency Preparedness, 1972).

Operation Foresight proved dramatically successful in reducing actual losses. It has been estimated that $250 million damage was prevented at a cost to federal, state, and local governments of about $36 million, a benefit–cost ratio of some 7:1. In addition, many of the temporary levees have since been incorporated into permanent flood protection works.

Simulations can also play a major role in preventing the illogical siting of new buildings since their negative effect on risk can be predicted before permits are issued. Similarly, they are useful in testing the benefits to be derived from strengthening building codes or altering by-laws which directly or indirectly influence risk. In this way, growth can be compared with its associated risk and decisions made on a rational basis. It is argued elsewhere in this volume that every municipality should prepare a plan for reconstruction long before any major damage is suffered. This precaution speeds recovery and reduces suffering. It can only be successful, however, if there is a clear understanding of where damage is likely to be the greatest and rebuilding, therefore, the most necessary. This information can best be provided by disaster simulations.

It should be noted that computer simulations are not without their drawbacks. These include the number and constraining effects of the underlying assumptions that must be made. There is often a lack of certain pertinent input data about various populations and infrastructures and of information on interrelationships among physical variables (White and Haas, 1975). Nevertheless, they can be quickly updated and have considerable merit.

Scenarios

Although computer simulations can be useful for predicting the nature of potential disasters in the near-term future, they lose much of their validity if they are intended to forecast over a longer time period. This is because their data base, and the assumed relationships between hazards and destruction on which they rest, can be made obsolete quickly by changes in building codes, land use patterns, and population distribution. One alternative approach to longer term disaster potential assessment involves the use of scenarios. The building of scenarios consists of considering the consequences of alternative assumptions about the future. Instead of trying to arrive at a "best estimate" of, for example, future fire or mercury pollution losses, several different possible alternatives can be considered in some detail. In this way a variety of futures can be discussed, actions and strategies necessary to attain them examined, the implied tradeoffs debated, and safety policies and objectives integrated with planning designed to satisfy a wide range of other social goals and needs (Sewell and Foster, 1976).

The scenario method attempts to set up a logical sequence of events, in order to ask how, if one starts from a set position, differing futures might evolve. As an exploration of futurity continues along one plausible pathway, a threshold or point of fundamental change is reached in the system which serves to initiate new directions to be explored. Scenarios provide a methodology for examining the consequences of specific decisions in a holistic setting. Their value lies in their ability to provide new insights into decisions which will stop, divert, or accelerate the evolution of a community at specific times. They assist in evaluating the validity of current goals and objectives and of major alternatives and their consequences. They, therefore, serve as tools for assessing all kinds of disaster mitigation policies (Ericksen, 1975).

A disaster scenario that has recently received a great deal of publicity was prepared at the request of the U.S. Senate Foreign Relations Committee for a floor debate on the Strategic Arms Limitation Treaty (Pincus, 1979). It is a fictional attempt to graphically describe what the world would be like after a nuclear war. The scenario envisages a massive Soviet attack that has been expected for several days. Tension between the super powers has prompted some people to flee major United States cities. As a result of the nuclear bombardment:

> Areas of the country such as the northeast corridor were reduced to a swath of burning rubble from north of Boston to south of Norfolk. Still there were some sections of the nation that were spared direct effects of blast and fire . . . Charlottesville . . . was not hit. . . . Momentarily [it was] immune to the disaster that had levelled the cities of the nation.

This scenario envisages that the Soviets have dropped some 4000 megatons on the United States and close to 100 million Americans have been killed outright. The United States counterattack has had a similarly devastating effect. The full consequences of the event are portrayed by graphically describing the effects of fallout on the Charlottesville survivors. In four days, this area receives a 2000-rem cumulative dose, enough to kill those who refuse to go to shelters and animals left outside. Refugees, many suffering from radiation sickness, pour into Charlottesville and overwhelm medical and shelter facilities. The former local government establishes an almost totalitarian rule, requiring identification cards, while rationing of flour, powdered milk, and lard is enforced. Private cars and tractors are outlawed because of an acute fuel shortage. Water is rationed since reservoirs are contaminated.

By the third and fourth weeks more people emerge from shelters and are forced to take in refugees from destroyed cities. Food shortages worsen and a black market develops. Radiation fatalities and hunger swell the death rate. Medical supplies have been all but exhausted and mass graves are established outside the city. Food riots begin as a grain truck arrives. Slowly national, state, and local government attempts to regroup. Young men are urged to return to the cities to aid in clean up; many refuse and are conscripted.

The financial system collapses as paper money is rejected and cottage industries develop making sandals, clothing, soap, and candles. Fall harvests are small, morale low, and in Charlottesville alone, several thousand people die in the first winter after the attack. As the scenario concludes, survival remains an uncertainty (Pincus, 1979).

The United States Defense Civil Preparedness Agency has developed a computerized system for generating scenarios for training purposes. Originally this focused on nuclear disasters (Dial-a-Scenario). It has since been expanded to include a capability for producing natural disaster scenarios. The computer system design incorporates the information for current operational checklists, service annexes, and historical reports and logs of previous actual and simulated disaster situations. Safety plan coordinators can specify the type of disaster, hazard level, and any special problems that they wish to be incorporated into the scenario. Once the type of disaster scenario has been denoted, a plausible sequence of events, spaced at realistic time intervals, is generated and can provide the basis for gaming or field exercises.

The basic types of disasters for which computerized scenarios can be produced in this manner are hurricanes, tornadoes, and earthquakes. Other adverse events can be included by using the set of special problems which include floods, toxic gases, fires, and accidents. Scenarios can be generated either in the time-shared or batch process modes. In the former, they are produced by remote terminal for use during a game or field exercise; in the latter, large numbers are generated at a computer processing center for later use (Baca, Bjorklund, and Laurino, 1975).

White and Haas (1975) have also pointed out the value of scenarios in generating alternative pasts. They allow, for example, the examination of what would have happened in Anchorage, Alaska, as a result of the 1964 earthquake if risk maps had been used to aid the location of major buildings. White and Haas argue that if a warning and preparedness program had been developed in the mid-1950s in Rapid City, the damage caused by the disastrous flash flood of 9 June 1972 would have been reduced by some 10%. A protection program of the type proposed in the 1940s would have brought about a far more drastic reduction in destruction (Figure 5.7).

To provide a degree of rigor and replicability in scenario building, White and Haas (1975) argue that a four-part model should be used. The first components in this heuristic device are the base conditions. These consist of a historical review of a hazard in a specific location over a definite time period, and the types of adjustment to it that were in fact considered, whether rejected or adopted. This historical record is handled as a social system whose dynamic components and their links form the essential characteristics of the object of the scenario, a natural or man-made hazard. Such a review provides a base from which to generate alternative pasts. Since it terminates in the present it can also provide a launching point for studies of potential futures. Used in this manner scenarios help to make explicit the reduction in damage and casualties that could have been, or will be, achieved if specific adjustments, such as better building codes or stricter air pollution standards, had been or are adopted.

The hazard system under review is also affected by external forces. These tend to be of two types. There are, for example, those that impinge upon the community from outside, such as national or state disaster mitigation policies, intergovernmental cost sharing for certain strategies, national insurance, or hazard mapping schemes. In addition the system also alters because of internal forces such as urban renewal, sprawl, park growth, and land use policies of many kinds. In the development of a scenario, assumptions must be made about how these factors will alter community risk and disaster potential. One very useful method for predicting such change is the Delphi technique; this is discussed in more detail later in this chapter.

The evolution of the hazard system over time is simulated by forecasting how the crucial factors identified in the historical review will be controlled or affected by such external variables. One of the chief objectives in writing scenarios is to illustrate the probable impacts of certain adjustments or strategies on other components of the hazard system.

Whereas such a historical overview is concerned with long-term approaches to decision making that emphasize thresholds and alternative directions, cross-sectional images focus attention on the immediate impact of a specific disaster agent. It is at such points that the efficiency of the community's safety program will be put to the test. (White and Haas, 1975). They are examined through the use of cross-sectional images. Given a particular scenario of policy adoption, what will happen if in 1987

1 Historical development assumes a 3½% per year growth rate.

2 Hypothetical flood plain land use regulatory program assumes a ½% per year decay rate.

3 Hypothetical protection program assumes a 80% reduction in damage potential and a 5% per year growth rate.

4 As in 3, but assumes in addition a 10% reduction in damages with an effective 2 hour warning period and effective emergency system.

5 Hypothetical warning and preparedness program assumes a reduction in damage potential of 10% per year.

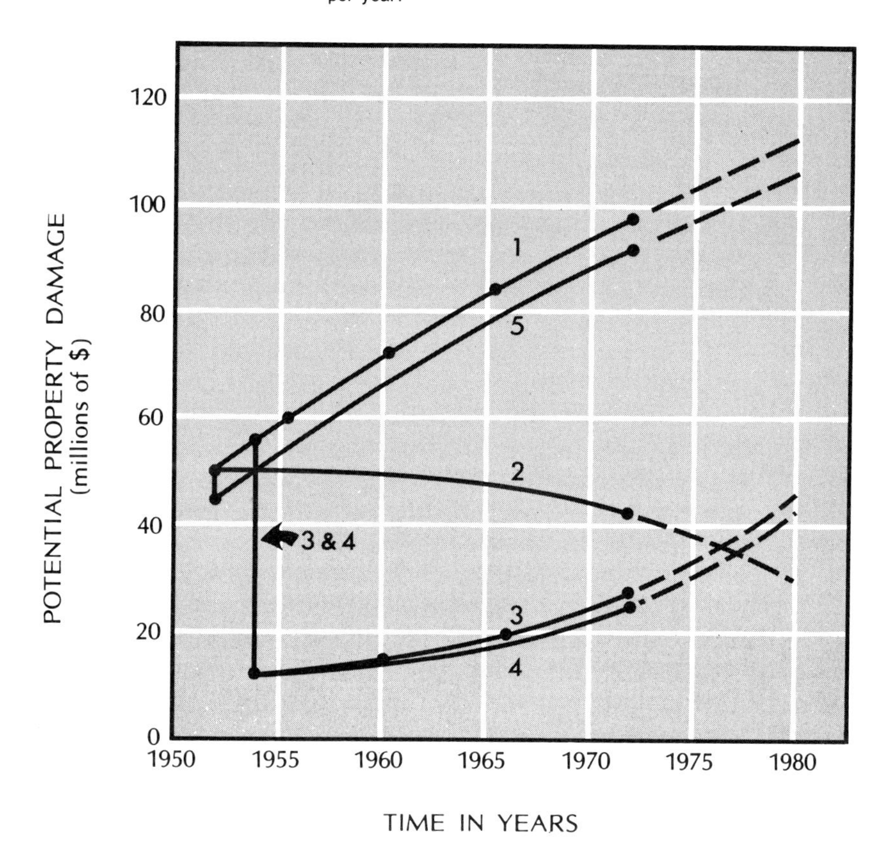

Figure 5.7. Potential property damage of alternative adjustments based upon the 9 June 1972 Rapid City flood (Source: White, G.F., and Haas, E., 1975, *Assessment of Research on Natural Hazards*. Copyright Massachusetts Institute of Technology, Cambridge, Massachusetts).

the spring freshet reaches 18.2 feet at a particular gauge? If an oil refinery doubles in size or production by 1989, what changes will occur in air quality and related illnesses? Such an approach allows the safety plan coordinator and his committee to estimate the likelihood that safety goals will be met if current policies are pursued. It also highlights what changes are required in strategies if this appears improbable.

There have been numerous discussions of the methodology and significance of scenario building. Those interested in applying the technique are referred to the works of Cole, Gershuny, and Miles (1978), Cole and Chichilnisky (1978), MacNulty (1977), and Smil (1977).

Safety plan coordinators can also use the scenario technique to gain public support for projects that were carried out despite opposition. An example of this approach can be seen in Rich's (1974) article on the Winnipeg floodway. This scheme involved more excavation than the Canadian section of the St. Lawrence Seaway and cost $61,276,000 to construct. Its chief aim is to divert flood waters from the Red River around Winnipeg. There was considerable resistance to its construction, yet in the spring of 1974 it prevented some $200 million in damage, at least as much may have also been avoided in early 1979. Safety plan coordinators should be aware that scenarios pointing out those areas that would have been destroyed and the degree of damage that would have occurred elsewhere if strategies had not been adopted can be valuable in gaining support for other disaster mitigation policies.

Delphi Technique

A weakness of both computer simulation and scenario building is that they are greatly influenced by external forces, alterations in the environment, and society-at-large that significantly modify the hazard system. Change is the only constant. For this reason, to increase the validity of such predictive techniques this flux must be accommodated as far as possible and its effects incorporated into these methodologies. One method of attempting to achieve this is the Delphi technique.

This methodology was developed by members of the Rand Corporation who described it in the following manner:

> Delphi is the name that has been applied to a technique designed to elicit opinions from a group with the aim of generating a group response. Delphi replaces direct confrontation and debate by a carefully planned, anonymous, orderly program of sequential individual interrogations usually conducted by questionnaires. The series of questionnaires is interspersed with feedback derived from the respondents. (Brown, Cochrane, and Dalkey, 1969)

There are three basic steps involved in applying the Delphic method. A group of knowledgeable individuals is selected and asked to participate in responding to several rounds of questioning. This is achieved by means of questionnaires, designed to eliminate psychological factors that influence results obtained in face-to-face interaction. Groups are often unduly affected by the outspokenness of dominant individuals and overt pressure toward conformity. This problem is avoided by the use of anonymous response through questionnaires. The topics discussed are those relevant to the qualifications of the group. In disaster-related Delphi exercises, the safety committee might approach a wide range of individuals, each likely to have insights into certain aspects of hazards or the nature of the social fabric. Normally each member of such a group would be asked to respond to a series of questions designed to solicit his or her views on when or whether certain events would probably occur, and if so, of what magnitude and significance were they likely to be. Table 3.1 might be used as part of such an exercise to gain insights in potential disasters the community faces.

Once the first round of questioning has been completed, responses are analyzed, and iteration and controlled feedback is employed to help participants to reevaluate and, if necessary, to revise earlier answers or to explain why extreme or unusual opinions are held (Smil, 1974). In this way, the questionnaire is reanswered and the group usually moves more toward a consensus. This process can be repeated several times until positions become solidified and all pertinent information has been assessed.

At this point, the group response is presented in statistical form, reducing the pressure toward conformity. Shared responsibility lessens inhibitions and each respondent's opinion influences the final outcome. Median-quartile evaluations are often used to give a clear view of both prevailing ideas and optimistic and pessimistic extremes (Linstone and Turoff, 1975).

This technique is essentially an exercise in opinion and value judgments (Smil, 1974). Best results are obtained from the participation of respondents who are very knowledgeable on the subject involved. It is particularly useful in the construction of scenarios since it allows the future development of external forces to be predicted by a group of experts, so mitigating the influence of personal bias.

In 1974, Smil published an account of a long-range forecasting study of the interrelationships between energy and the environment. This research project involved the use of the Delphi method to gain a world perspective well into the twenty-first century. His results were based on the responses of 40 participants from North America, Europe, and Japan. Table 5.4 is a summary of the views expressed by this group on the probabilities of the occurrence of various energy-related man-made disasters in the 1970s. It is now possible to compare these forecasts with reality. Table 5.5 presents a ranking of the significance of energy-related hazards on a global scale. From this overview it can be seen that the experts polled considered internal combustion engine emissions to be the most important threat, followed by sulfur dioxide resulting from power generation.

The Delphi technique has been used in a wide variety of situations as diverse as predicting the likely demand for a new ghost story (Anon., 1977) to forecasting China's future (Smil, 1977). Sackman (1976) es-

Table 5.4. Probabilities of Environmental Episodes in the 1970s

Episode	Lower quartile (%)	Median (%)	Upper quartile (%)	Mode (%)
1. Severe urban air pollution episode lasting several days with significant consequences	40	90	100	100
2. Widespread failure of power supply in populated industrial region, lasting several hours	50	70	100	100
3. Catastrophe of fully loaded jumbo tanker (over 100,000 dwt) and spill of crude oil in the open sea	25	70	95	100
4. Serious oil spill from offshore drilling operation causing ecological disturbance over a large area	20	50	75	20
5. Radioactive contamination of environment outside of a reactor building caused by failure of nuclear plant protective systems	5	5	10	5

Source: Smil (1974).

timates that over 1000 Delphi studies have already appeared. Despite its many advantages as a technique for stimulating the future of local hazard systems, potential users should be aware of various criticisms levied against it. These include biases in the selection of the panel of respondents, the effects of group suggestion, and the possibility of sloppy execution by experts in a hurry (Sackman, 1976). Nevertheless, the Delphi

Table 5.5. Consensus on the Priorities of Energy–Environment Problems

Rank	Item	Percent weight in the Final round	Percent weight in the Middle round
1.	Internal combustion engine emissions	7.3	7.4
2.	Sulfer dioxide from power generations	6.9	7.1
3.	Particulate matter from combustion	6.4	6.7
4.	Water pollution by energy production systems	5.9	6.1
5.	Thermal pollution by nuclear power plants	5.9	6.1
6.	Nitrogen oxides from power generation	5.8	5.6
7.	Accidental oil spills	5.6	5.4
8.	Accumulation of wastes	5.0	3.9
9.	Domestic heating (waste of fuels, air pollution)	4.9	4.4
10.	Radioactive wastes and emissions	4.8	5.1
11.	Thermal pollution by fossil fuel fired power plants	4.6	4.1
12.	Irresponsibility of energy management	4.6	4.6
13.	Coal mining environmental disturbances	4.4	5.0
14.	Carbon dioxide in upper atmosphere	3.5	3.1
15.	Visual pollution by overhead lines	3.0	3.5
16.	Concentration of pollution (economies of scale)	2.9	2.6
17.	Fossil fuels for combustion instead of chemical source	2.9	2.7
18.	Conservationists	2.7	2.8
19.	Export of modern technology to underdeveloped countries	2.6	2.6
20.	Miner's hazard in procuring fossil and nuclear fuels	2.2	1.9
21.	Ecological and aesthetic impact of large dams	2.1	2.5
22.	Carcinogenic effects of aromatic petroleum products	1.7	1.8
23.	Land use by power line rights-of-way	1.6	1.8
24.	Fog from wet cooling towers	1.3	1.7
25.	Earthquake danger to nuclear power plants	1.2	1.5

Source: Smil (1974).

technique has considerable potential for wider use as a tool for assisting in the design of warning systems and the compilation of disaster and reconstruction plans. Local experts might be asked by the safety committee to rank threats in order of significance; assign probabilities to the occurrence of different types of disasters in the region; set priorities for strategy implementation; and predict changes in the social fabric or environment that are likely to influence risk. Apart from the value of the information collected, such participation may encourage acceptance of disaster mitigation strategies and help to increase community awareness.

The next logical step in the use of modeling to reduce the impact of hazards involves a simulated response to the threat itself. That is, not only is the expected loss predicted, but so too is the reaction of the victims and officials and agencies that must accept responsibility to meet the demands it creates. Some of this information can be generated by game simulations.

Game Simulations

It is not adequate merely to simulate alternative futures. Safety committees and those involved in reducing disaster losses must also seek to influence the selection of a less hazardous futurity by the community, and to learn how to make optimum responses to hazard threats. Game simulations are a useful tool for attempting such tasks. As McLuhan (1964) has pointed out: "games like institutions are extensions of social man and of the body politic as technologies are extensions of the animal."

By stressing decision making, game simulations seek to model and increase understanding of the dynamic processes of society and the ways in which its elements actually function (Wood, 1973). In applying this technique, participants assume the role of players and act out a model of a real world situation. In disaster gaming, for example, players would normally pose as victims, medical personnel, disaster planners, city officials, police and fire chiefs and their staffs, and a wide variety of the other groups directly involved in responding to disaster.

Gaming simulations are popular instructional techniques, since few people can resist being asked to project themselves from formal learning procedures into situations of systematically developed involvement. Such simulations appear to have a particularly valid role to play in disaster mitigation research since one cannot simply learn from adverse experience. When a community actually suffers a disaster it is too late to use this experience to pinpoint deficiences in equipment or planning procedures or to discover errors in personnel training. The costs of such inefficiences are too high. It is far more realistic to repeatedly model disaster situations through game simulations and to have those with the responsibility of responding to catastrophe act out their own roles and those of others, to discover weaknesses in preparedness.

Such modeling of real world situations through gaming has certain specific advantages:

> As a form of simulation model, gaming is similar to other simulations where the major purpose is to understand dynamic processes. While the approach emphasizes process, it attempts to comprehend human activities as the products of indeterministic forces which can lead to any one of a possible range of outcomes, and is in essence a probabilistic approach. Where gaming simulation differs from other methods is that it attempts to provide experience of a 'real world' decision-making situation, where goals have to be formulated, problems evaluated and judgement exercised. (Wood, 1973)

The main components of game simulations are arranged as a structured learning system in Figure 5.8. This demonstrates the links between problems and players who seek to understand them. It illustrates that there are eight main components of any such simulation procedure (Wood, 1973):

1. Initial problem definition and the scale of the game
2. Umpires and players, involved individuals, groups, power structure, mores, laws, goals
3. Information given and generated
4. Play: the occurrence of events, chance factors
5. Temporal development
6. Accounting procedures
7. Techniques of communication
8. Evaluation

Problem Definition

The first issue to be resolved is that of the game's primary purpose. Why is the simulation being conducted? Organizers must be very specific about their rationale for the exercise. In disaster mitigation gaming this will normally be to determine likely and optimum community responses to the impact of a particular man-made or natural disaster agent. The hazard selected for simulation and the frequency, intensity, and areal extent of its probable impact might be predetermined by the organizers through the earlier use of scale models, computer simulations, scenarios, or the Delphi method. Such techniques will also assist in defining realistic areal boundaries for the game, that is, to limit the geographical extent of the disaster. In addition, those groups who would be likely to play a significant role as either victims or participants in disaster plan implementation must be defined. While some organizational compromise may be necessary to keep the simulation manageable, representatives of as many of such groups as possible should be encouraged to become directly involved.

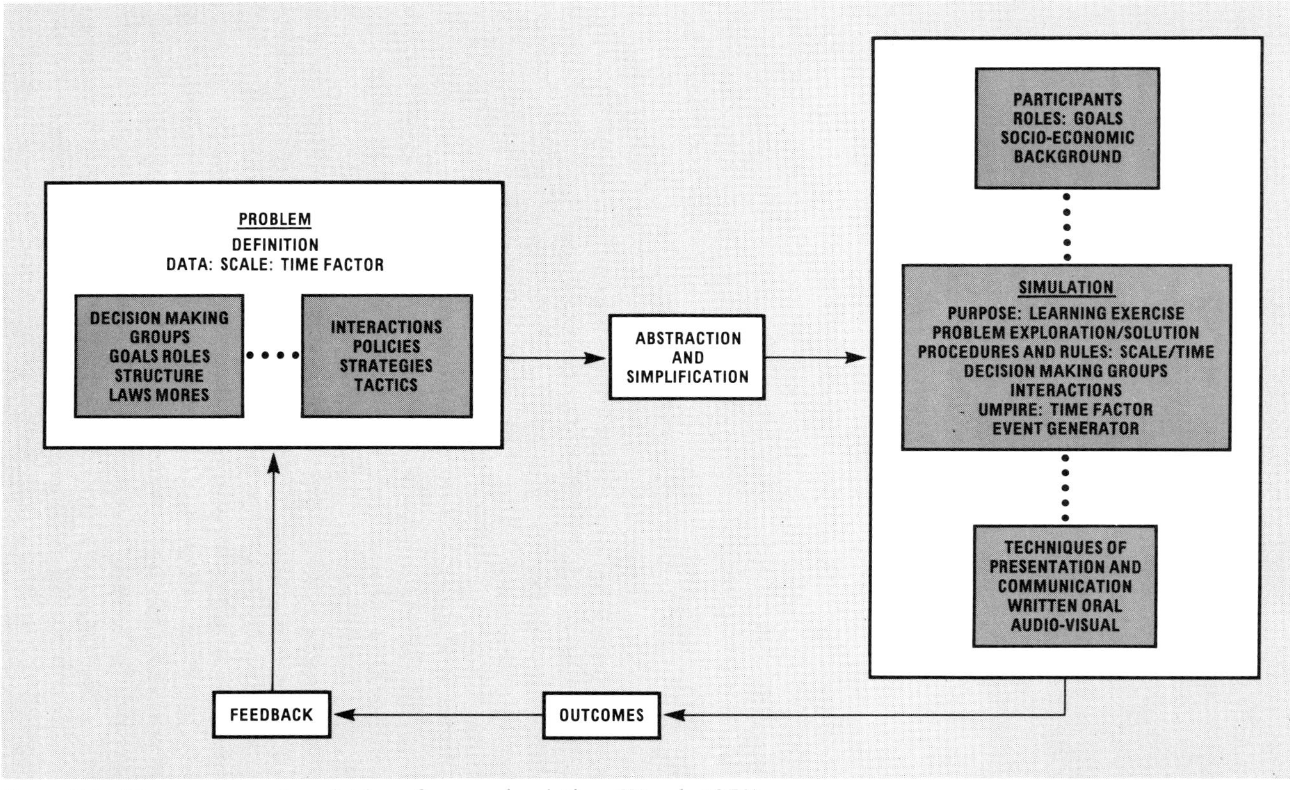

Figure 5.8. The component variables of game simulation (Wood, 1973).

Umpires and Players

Games can be controlled by either an umpire or through group consensus. Where there is a lack of experience among the players or if they are unfamiliar with the subject of the game, an umpire may be needed. Where players are better prepared there may be devolution of power. Ideally in simulating disaster situations, power and responsibility should be distributed according to the actual rank or influence of the individual or group concerned. For example, the mayor, civil defense coordinator, and senior medical officers of health would normally play key roles. In certain man-made disasters scientists would be expected to provide expert guidance.

Players must have clearly defined roles and goals. In the case of groups, leaders may be elected to make final decisions. Invariably the goals of the competitors and the intent of the game are inextricably interlocked. These should be delineated either by the umpire or by the players as the game proceeds. In disaster gaming most players will seek to minimize losses of life and property, although certain groups such as looters and terrorists may attempt to satisfy antisocial personal objectives.

Players' strategies or choices can either be stated or unstated. They must be designed to assist in goal achievement. In the case of disaster simulations of this type, strategies might be designed to ensure safety goals are met. These may vary from extreme conservatism (minimizing loss) to high risk taking (maximizing gains). In complete contrast to extremely competitive situations such as land development games, disaster simulations would normally emphasize compromise or cooperation (Patchen, 1970). Most games are designed to model the distribution of power between groups, either in the form of votes, money, or resources. Disaster gaming would tend to emphasize the impact of hazards on the distribution of the two latter surrogates of power.

Information

Two types of information are normally available to players: primary information which is made available at the onset and secondary data generated during the game itself. In disaster situations the collection and analysis of data is a significant task. For this reason, gaming in this area should generally involve the provision of relatively little primary information. Additional data might be collected and systematically released to the players as the game progresses. Groups representing the operators of disaster warning systems, the media, or other data-generating sources may play this role. The information provided should be as realistic as possible and the value of the game is greatly improved if data simulated by computer, scale models, or Delphi techniques are used for this purpose. Furthermore, deliberately false information can be incorporated to represent error. Information overload, so common in disaster situations, might also be reproduced. Gaming can be conducted with complete or limited communication between groups to demonstrate the value of a central command post and interagency coordination.

Play

Once the purpose, sophistication, and scale of the problem have been defined, the information input procedure decided and the number and role of groups specified, the game can begin. This is usually arranged and coordinated by the umpire and players according to established rules. In disaster simulations these would represent legislation, economic and technical feasibility, and social acceptability. If a community disaster plan already exists this can be used to provide the framework to control the progression of the game.

The Chance Factor

Reality is extremely complex. Rare and completely unpredicted events may occur, thus complicating or compounding disaster. A terrorist group may decide to use the confusion created by a flood to sabotage a nuclear power station or poison the city's water supply system. A dam may collapse as the result of an earthquake. In many games, chance factors that may influence its outcome are tabulated and randomly injected into the simulation. The actual time of input may be selected by the use of random numbers or from shuffled cards. Chance factors are designed to train players to react quickly and logically to the unexpected. It is possible for one group of players to act as terrorists or looters with goals that conflict with those of other participants. They will then be responsible for injecting such chance factors into the game.

Time

A common operational procedure involves dividing the game into rounds to simulate temporal development (Wood, 1973). In disaster gaming logical breaks would seem to occur between the preimpact and impact phases, the emergency period, and the restoration and reconstruction stages (see Chapter 8). The speed of simulation is also another important variable. In the early stages of the disaster process, time is usually in very short supply. For this reason, players should be placed under stress to respond quickly, even when data are very far from complete.

Accounting

A fundamental component of any game is an accounting procedure to evaluate performance. In this way improvements can be monitored if the simulation is repeated. In most situations, the traditional bookkeeping is the accounting scheme, showing performance in terms of credits, debits, profits, and loss (Walford, 1969). Other accounting procedures involve the use of the cost–benefit method or variations of it such as the balance sheet of development. In some cases a "goals achievement matrix" approach is used (Hill, 1968). In hazard situations the cost of operations is often relatively immaterial. Success might be measured in terms of casualties prevented, or stress or damage reduction.

Techniques of Communication

Information may be provided to players by oral communication. In real disaster situations, however, other techniques may be needed to supply data and to link, integrate, and involve players. Slides, films, videotapes, and audiotapes of past disasters can be used to heighten realism. Computers can be utilized to provide information or to aid in statistical processing and decision making. Radio communications and telephone links can be used to simulate input from field observers.

Evaluation

The value of such gaming lies in the understanding it gives to the players of the problems they would face in real life disasters. Decisions made during the simulation should, therefore, be evaluated at leisure to see whether in fact they were the optimum response. In this way, game simulations become a painless trial and error approach to solving the unique problems disaster poses in a community. Naturally the lessons learned must eventually be incorporated into the operation of warning systems, disaster plans and reconstruction designs.

Gaming: An Overview

There are many urban and rural development games that can be ubiquitously applied. Reference to a substantial number of these can be found in publications by Duke (1964), Walford (1969), Taylor (1971), Kibel (1972), and Wood (1973). In addition, a fairly comprehensive listing of the available literature is available in *Gaming Techniques for City Planning: A Bibliography*, published by the Council of Planning Librarians (Thornton, 1971).

Games specifically designed to simulate hazard impact include POLLUTION produced by Abt. Associates Inc. for Wellesley, Massachusetts, public schools; APEX (Air Pollution Exercise) created by Duke *et al.* (1969); and THE DISASTER GAME developed by Inbar (1965). Ideally, each game should be specifically designed for a particular community. Wood (1973), for example, has developed a model which simulates an oil tanker disaster in the Strait of Juan de Fuca.

On a larger scale the Canadian federal government has been conducting gaming to test contingency plans for an oil spill in the Beaufort Sea, in the southeastern corner of the Arctic Ocean. There has already been exploratory offshore drilling in this harsh and delicate environment (Mansfield and Hoffman, 1979). Exercise BREX III was carried out in 1978 to test Canadian ability to deal with a serious oil spill in this area. It involved a combination of scenario development, resource and exercise control staffs, and simulations of development and radio/landline communications through eight operating centers connected by special telephones. Approximately 80 participants from government and industry took part in the game; it lasted $2\frac{1}{2}$ days in the Territorial Government offices, Whitehorse, Yukon.

The scenario involved a hypothetical large blowout which resulted in oil on beaches, open-water slicks, and threatened fish and wildlife habitats. Various organizational groups were staffed with players while a separate group of control staff interacted with players as the scenario developed. These controllers responded to players' directions and added appropriate new problems as the situation developed. Players were forced to assess the situation, react according to their plans and responsibilities, define priorities and tasks, allocate resources, and issue orders and communicate movement. The problems injected became more difficult to solve as the players got the feel of the exercise. Once the game was over the plan for responding to such a spill was reevaluated. The main lessons learned were the need for a better defined command structure, coordination, and better logistics/administrative support arrangements (Mansfield and Hoffman, 1979).

Probably one of the most comprehensive of all disaster games is ATLANTIS, the Simulation Exercise in Disaster Relief Management developed by Ritchie and the International Systems Corporation of Lancaster (G. N. Ritchie, 1978, personal communication). This game is designed to provide a realistic simulation in which the techniques of resource management in disaster can be practiced. It last for between two to four days and requires up to 15 players acting singly or in pairs to represent national ministries on a National Disaster Emergency Committee. The exercise follows a pattern of free play with participants being responsible for policy decisions and plans for executive action.

The simulation itself is based on a computer program in which a series of models interact to represent the production and consumption of food, water, fuel, power, and manufacturing processes. Deaths, injuries, and destruction of the infrastructure and the means of relief and repair and the transportation required to deploy these are also simulated.

Throughout the game the Exercise Directing Staff and the computer play the role of the society of ATLANTIS, its people, and of the world beyond. The Directing Staff reacts to the players' decisions and plans, while the computer rapidly analyzes their implications. The exercise is designed to examine three phases of disaster. In the first the players seek information concerning the situation, assess relief needs, and gain control of resources and the capabilities needed to use them most effectively. In the second phase, they react to the emergency situation and deploy immediately available resources. In the third and final phase the reception, receipt, and distribution of foreign relief resources are handled.

ATLANTIS was specifically designed as a training tool in disaster management for administrators in developing countries. It has been used, however, with considerable success by officials with emergency services training in Britain and in NATO (G. N. Ritchie, 1978, personal communication).

There are also several games designed by military strategists to model combat. Since the war is a form of man-made disaster these are also relevant here. They include a simulation of the Vietnamese conflict (Macrae and Smoker, 1967) and war in general (Featherstone, 1962).

Despite the many advantages of disaster gaming, this technique still lacks the total involvement that catastrophe inevitably brings. Many actual difficulties may be overlooked in the simulation. Gaming and the learning experience it provides should, therefore, be taken one stage further by the use of field exercises. These seek to model the disaster experience outside the classroom and the office.

Field Exercises

Unfortunately, all too often the community disaster plan is designed in isolation by a committee, which then circulates a report to the individuals and agencies designated to respond in the event of an actual emergency. It is commonplace for such documents to be read briefly and then filed and virtually forgotten. As a result, disaster plans too often remain at the theoretical level and are not rehearsed in simulation exercises. This lack of practice allows gaps, inconsistencies, and errors to go indetected. When disaster actually strikes, weaknesses in planning become immediately apparent and crucial time and personnel must be used to compensate for such inadequate arrangements. This lack of rehearsal also leads to unfamiliarity with existing procedures and facilities by both the officials responsible for their ultimate implementation and by the public in general.

To avoid this error the safety program coordinator and his committee can organize field exercises to simulate disaster. These should be based upon the more likely emergencies that the community might actually have to respond to. To ensure that they are as realistic as possible field exercises should be preceded by the application of some of the techniques previously described. The use of models and computer simulations, for example, might provide data on the scale of probable damage and the number and distribution of potential casualties. Scenarios and gaming can help to predict social response. This information is invaluable in making certain that disaster field exercises realistically portray the impact of particular hazards.

The timing of simulation field exercises is a significant factor to be considered. It is not realistic to advise the medical staff, ambulance drivers, disaster officials, and others that a particular type of emergency situation will occur on a specified date at a given time. Most hazards are simply not that accommodating. Disasters do not necessarily take place on sunny days at 10:30 A.M. Emergency response teams should be just as prepared to take part in simulations enacted at 2:30 A.M. in freezing temperatures or during a rainstorm. Forewarning those involved almost inevitably leads to the creation of a false veneer of readiness that interferes with realistic evaluation.

It is important to provide participants with as an authentic disaster situation as possible to respond to. Buildings scheduled for demolition might be set on fire or destroyed. Film and audio equipment can be used to broadcast the sights and noises of actual disaster. Wreckage can be spread around the site. Communication facilities, equipment, or other materials that are likely to have been damaged can be ruled off limits to the emergency response team. In most communities there are many individuals or groups who are willing to assist with the staging of a disaster field exercise. Since such simulations are, in effect, a piece of theater, produced not to entertain but to instruct, local dramatic societies, college and university theater departments, and other thespians might be approached to play the roles of victims or their families. Service groups or boys' and girls' clubs might also act these parts. Each such participant should be provided with a script to memorize which permits officials responding to the simulated disaster to deduce the nature of their probable injuries, if any, the possible location of missing persons, and the presence of secondary threats. The data provided might be incomplete or accidentally misleading.

If possible, the cooperation of the local news media should be obtained. They can play two very significant roles. Producers and directors can assist in the actual staging of the "disaster" while their colleagues can inform the public of the lessons learned from the exercise. In this way community awareness and readiness during actual emergencies can be increased. Insurance should be taken out to cover the organizers in case of accidental injury to participants.

The purpose of field exercises is to improve the operation of the disaster response system so that it performs more effectively in actual emergencies (Davis, 1975). For this reason, it is essential that every facet of the simulation be carefully recorded for later evaluation. This process of hindsight review is likely to be most successful if detailed records are kept of the wording and time of arrival of all related data, including warning messages and the system's response to them. The task of data collection should be given to a team of observers who are in no way related to the emergency response program. They should be provided with log books, tape recorders, and videotape equipment and be permitted access to all and every stage of the simulation.

Once the simulated disaster has passed, the observers should be given the opportunity to debrief those involved. They should then retire and quickly produce a report which fairly highlights the strengths and weaknesses of the emergency response team's efforts. Once this has been made available to all concerned, it should be discussed in detail and efforts made to remedy apparent weaknesses. It can also be used as the basis for a gaming exercise which can help to improve the emergency team's readiness before the next field exercise. Regardless of the criticisms included in such an evaluation, it should be remembered that this document can only lead to improved performance.

Before a community undertakes a disaster field exercise for the first time, it is recommended that the safety committee review simulation evaluations from other areas. Many of these are available. For example, there is the report on Operation Crash, a simulated disaster based on the pretense of a DC-9 aircraft flying into the Midtown Plaza Building in Saskatoon. Its major aim was to test the validity of the University Hospital's disaster plan (Godsalve, 1973). Problems were encountered with the tagging of victims (the writing was so bad that tags could not be read); the messenger service was inadequate as were telephone communications and the speakers used to sound alerts. In summary, many of the difficulties encountered in implementing the hospital disaster plan were practical. These would have been very difficult to predict in a gaming situation but could easily be corrected once they had been identified through field exercises.

Another example of such a community disaster exercise took place on 30 October 1974 in London, Ontario. The drama unfolded when an RCMP corporal telphoned city police to say that a DC-9 jetliner en route from Cleveland had disappeared off radar screens (Sutherland, 1975). It was later learned that the imaginary plane had crashed in the city's core. London's disaster plan and that of five city hospitals were activated. A number of prearranged callers jammed the police switchboard with crash details. During the simulation all calls relating to the imaginary event were prefixed with the phrase "Exercise London." In this way if a real disaster had occurred at the same time, officials could have distinguished between the two. The exercise involved 150 casualties, 27 of whom were fatalities.

Problems discovered included an inability of some hospitals to deal simultaneously with large numbers of critically injured, all requiring surgery. The transfer of vital information and instructions between and within hospitals was also found to be inadequate. Hospital switchboards became clogged with calls from supposed relatives seeking patient condition reports. As with the Saskatoon exercise, illegible handwriting on patient tags caused unnecessary and, under real disaster conditions, possibly fatal delays. Color-coded tags might have been used to denote the seriousness of injuries.

One of the most comprehensive disaster field exercises ever staged, Operation Nuwax-79, was enacted in April 1979. This simulation took four years to plan and cost $1.6 million to prepare and execute. It was designed to rehearse the United States' ability to deal with the crash of a nuclear armed warplane. Responding to the code alert "Broken Arrow Mayday," several cargo jets carrying radiation specialists and monitoring and response equipment rushed to Jackass Flats, 100 miles north of Las Vegas. It took four hours to locate the actual crash scene, marked by aircraft wreckage including pieces of actual nuclear warheads, three damaged and three destroyed by fire on impact. Six humanlike forms, representing dead crew members, lay near the debris. The site was indeed radioactive, being sprayed previously with a solution containing radium 223 to reproduce the effects of plutonium. This exercise, which lasted approximately one week, involved combining the efforts of the United States Air Force, Army, Navy, and civilians to handle problems of contaminated bodies, wind scattered radiation, classified nuclear warheads, and radioactive wreckage. If the exercise had been an actual disaster the cleanup process would have taken several months.

The value of such field exercises was demonstrated in Topeka, Kansas, where yearly simulations are held to coincide with the start of the tornado season. When the community was actually damaged in 1966 its disaster plans and the personnel responsible for their implementation proved to be very effective, due in no small part to their constant rehearsal (Parr, 1969). Even when the field exercises have not dealt with the type of disaster that actually later befalls a community, they can still be of great value in improving response. In October 1974 a research team from Ottawa's Carleton University visited Sydney, Nova Scotia, following a destructive wind storm. Its purpose was to evaluate the community's response to this disaster. The evaluation proved very positive and included the assertion that reaction had probably been so effective because of a field exercise held in the community some 16 months before (Scanlon and Jefferson, 1975). This was the case despite the fact that the actual disaster experienced was a windstorm while the simulated event was the collapse of a wing of the Sydney Academy, a local high school. Numerous differences occurred between simulated and actual casualties, damage, and threats. Nevertheless, it was concluded that the field simulation had been of great value for two reasons. First, it gave some persons direct experience of the problem of handling a disaster, especially with decision making under stress and with emergency communications. Second, Scanlon and Jefferson saw such field exercises as being a major stimulus to community awareness of the need for a disaster response. Many of those who took part in the field simulation were very willing to play similar, or more responsible roles, in the actual disaster. Having seen the value of disaster exercises the city appears ready to hold them regularly.

Conclusion

Simulation of futurity permits a society far greater choice in selecting its own future than would otherwise be the case. Since the avoidance of disaster, or at least reduction of its magnitude, is a ubiquitous aim, disaster modeling should normally play a significant role in any community debate about alternative futures. There appear to be a series of logical steps in the application of such aids. The first of these involves the use of scale and computer models to determine the physical characteristics of particular disaster agents, such as floods, fires, air pollutants, or disease. These permit predictions to be made of when, where and how such hazards are likely to seriously interfere with the normal functioning of the community. Computer simulations including data on the nature of society can then be used to forecast, with considerable accuracy, resultant casualties and damage. More distant threats can be examined by less precise methods, such as the development of scenarios and the application of Delphi techniques. It is not enough merely to predict, a community must learn to react. Gaming and field exercises teach the emergency response team and the community in general what to expect if disaster strikes, and encourage them to function effectively. Simulated mistakes are far less costly than the real thing.

The modeling of disaster should also influence communities in a more profound way. It permits planners and legislators to understand how land use decisions they routinely make influence risk and, therefore, the potential for future disaster. It may lead to changes in policy and a greater concern about safety through design. As the tools for predicting and mitigating disaster improve, so does the validity of the truism that every community experiences the losses it deserves.

References

Allen, J. 1947. *Scale Models in Hydraulic Engineering*. Cited in Chorley, R. J., and P. Haggett (eds.) 1967. *Models in Geography*. Methuen, London. p. 145.

Anon. 1977. Delphi and short-term market forecasting. *Futures*, **8(2)**:171–172.

Appleby, F. V. 1956. Run-off dynamics—A heat conducting analogue of storage flow in channel networks. *International Association of Hydrology*, Publication No. 38 (Rome), **3**:338–348.

Baca, J., P. B. Bjorklund, and R. K. Laurino. 1975. *Natural Disaster Scenario Generators*. Report prepared for U.S. Defense Civil Preparedness Agency, Washington, D.C., 35 pp.

Beasley, R. P. 1972. *Erosion and Sediment Pollution Control.* Iowa State University Press, Ames, Iowa, 320 pp.

Bishop, A. W., J. N. Hutchinson, A. D. M. Penman, and H. E. Evans. 1969. *Geotechnical Investigation into the Causes and Circumstances of the Disaster on 21st October, 1966. In: Selection of Technical Reports Submitted to the Aberfan Tribunal.* Her Majesty's Stationery Office, London, pp. 1–80

Brown, B., S. Cochrane, and N. Dalkey. 1969. *The Delphi Method II: Structure of Experiments.* Rand Corp., Santa Monica, California, RM-5957-PR, 131 pp.

Bryan, R. 1967. On rainfall simulation in field methods for the study of slope and fluvial processes. Commission on Slopes and Submission on Fluvial Dynamics of the Commission on Applied Geomorphology, International Geographical Union. *Revue Geomorph. Dynamique,* **17(4)**:146–188.

Chapon, J. 1961. Advantages of the 'historical, method in mobile-bed model testing: Application to the Seine estuary. *The Dock and Harbour Authority,* **42**:489.

Chorley, R. J., and P. Haggett. (eds.) 1967. *Models in Geography.* Methuen, London. 816 pp.

Chorley, R. J., and B. A. Kennedy. 1971. *Physical Geography: A Systems Approach.* Prentice-Hall International, London, 370 pp.

Coates, V. T. 1972. Technology and public policy: The process of technology assessment in the federal goverment. *Program of Policy Studies in Science and Technology,* **1.** George Washington University, Washington, D.C.

Cochrane, H. C., J. E. Haas, M. J. Bowden, and R. W. Kates. 1974. *Social Science Perspectives on the Coming San Francisco Earthquake: Economic Impact, Prediction and Reconstruction.* University of Colorado, Institute of Behavioral Science, Natural Hazards Research Working Paper No. 25, Boulder, Colorado, 81 pp.

Cole, S. and G. Chichilnisky. 1978. Modelling with scenarios: Technology in north-south development. *Futures,* **10(4)**:303–321.

Cole, S., J. Gershuny, and I. Miles. 1978. Scenarios of world development. *Futures,* **10(1)**:3–20.

Davidson, R. D., and R. W. Whalin. 1974. *Potential Landslide-generated Water Waves, Libby Dam and Lake Koocanusa, Montana: Hydraulic Model Investigation.* Final Report. United States Army Engineering Waterways Experiment Station, Vicksburg, Mississippi. Technical Report TR-H-74-15, 123 pp.

Davis, J. E. 1975. After the storm is too late. *Emergency Planning Digest,* **2(3)**:7–9.

Duke, R. D. 1964. *Gaming Simulation in Urban Research.* Institute for Community Development and Services, Michigan State University, East Lansing, Michigan.

Duke, R., *et al.* 1969. *APEX (Air Pollution Exercise).* University of Michigan, Mimeograph.

Einstein, H. A., and J. A. Harder. 1961. Electric analog model of a tidal estuary. *Transactions of the American Society of Civil Engineers,* **126(4)**:855

Ericksen, N. J. 1975. *Scenario Methodology in Natural Hazards Research.* Institute of Behavioural Science, The University of Colorado, Program on Technology, Environment and Man, Monograph NSF-RA-E-75-010.

Featherstone, D. F. 1962. *War Games.* Stanley, London.

Foster, H. D., and R. F. Carey. 1976. The simulation of earthquake damage. *In:* Harold D. Foster (Ed.), *Victoria Physical Environment and Development, 12,* Western Geographical Series. University of Victoria, Victoria, B.C., 334 pp.

Friedman, D. G. 1969. Computer simulation of the earthquake hazard. *Proceedings of Geological Hazards and Public Problems Conference,* May 27–28, 1969. Region Seven, Office of Emergency Preparedness, Santa Rosa, California, pp. 153–181.

Friedman, D. G. 1973. *Computer Simulation of Natural Hazard Effects.* The Travelers Insurance Company, Hartford, Connecticut.

Friedman, D. G., and R. S. Roy. 1966. *Simulation of Total Flood Loss Experience on Dwellings in Inland and Coastal Flood Plains.* Report prepared for the U.S. Department of Housing and Urban Development. The Travelers Insurance Company, Hartford, Connecticut.

Gibson, M. E. 1976. San Francisco Bay under a roof. *Sea Frontiers,* **22(3)**:166–171.

Godsalve, W. H. L. 1973. *Final Report 'Operation Crash.'* Saskatoon Emergency Measures Organization, Saskatoon, Saskatchewan, 30 pp.

Harder, J. A. 1963. Analog models for flood control systems. *Transactions of the American Society of Civil Engineers,* **128(1)**:993

Henchey, N. 1978. Making sense of future studies. *Alternatives: Perspectives on Society and Environment,***7(2)**:24–28.

Hill, M. 1968. A goals achievement matrix for evaluating alternative plans. *Journal of the American Institute of Planners,* **34**:19–29

Inbar, M. 1965. *Simulation of Social Process: The Disaster Game.* The Johns Hopkins University, Department of Social Relations, Baltimore, Maryland.

Japanese National Research Center for Disaster Prevention. 1971. Diasaster prevention—Japan. *Emergency Measures Organization National Digest,* **11(4)**:27–29.

Kaplan, M. 1971-1972. Actuarial aspects of flood and earthquake insurance. *Proceedings of the Conference of Actuaries in Public Practice,* Chicago, **21**:474–511.

Kessler, E. 1977. Thunderstorms and hurricanes. *McGraw-Hill Yearbook Science and Technology.* McGraw-Hill, New York.

Kibel, B. M. 1972. *Simulation of the Urban Environment.* Association of American Geographers, Commission on College Geography, Technical Paper Series No. 5.

Linstone, H., and M. Turoff (Eds.). 1975. *The Delphi Method.* Addison Wesley, Reading, Massachusetts, 620 pp.

MacNulty, C. A. R. 1977. Scenario development for corporate planning. *Futures,* **9(2)**:128–138.

Macrae, H., and P. Smoker. 1967. A Vietnam simulation: A report on the Canadian/English joint project. *Journal of Peace Research,* **4**:1–25.

Mansfield, B., and J. Hoffman. 1979. Government contingency plans for the Beaufort Sea. *Emergency Planning Digest,* **6(1)**:15–19.

McIntyre, R. J. 1979. Analytic models for West Coast storm surges, with application to events of January 1976. *Applied Mathematical Modelling*, **3**:89–95.

McLuhan, M. 1964. *Understanding Media: The Extension of Man*. McGraw-Hall, New York.

Milne, W. G., and A. G. Davenport. 1969. Distribution of earthquake risk in Canada. *Bulletin of the Seismological Society of America*, **59**:729–754.

Panuzio, F. L. 1968. The Hudson River model. *Hudson River Ecology Symposium, 1966, Tuxedo, New York*. Hudson River Valley Commission, New York, pp. 83–113.

Parr, A. R. 1969. A brief on disaster plans. *Emergency Measures Organization National Digest*, **9(4)**:13–15.

Patchen, M. 1970. Models of cooperation and conflict. *Conflict Resolution*, **XIV(1)**:389–407.

Pincus, W. 1979. N-Blast ... then the horror begins. *Victoria Times*, May 29, pp. 1–2.

Rich, S. G. 1974. Floodway prevents 1974 Winnipeg flood. *The Canadian Geographical Magazine* **88(6)**:20–29.

Rinehart, W., S. T. Algermissen, and M. Gibbons. 1976. *Estimation of Earthquake Losses to Single Family Dwellings*. United States Geological Survey PB-251-677.

Sackman, H. 1976. A sceptic at the oracle. *Futures*, **8(5)**:444–446.

Scanlon, J., and J. Jefferson. 1975. The Sydney simulation. *Emergency Planning Digest*, **2(6)**:2–7.

Scanlon, J., J. Jefferson, and D. Sproat. 1977. Mud slide in Port Alice, Canada. *Ekistics/Oikietikh: The Problems and Science of Human Settlements*, **44(260)**:27–31.

Schmidt, B. L., W. D. Shrader, and W. C. Moldenhauer. 1964. Relative erodibility of three loess-deprived soils in southwestern Iowa. *Proceedings of the Soil Science Society of America*, **28(4)**:570–574.

Sewell, W. R. D., and H. D. Foster. 1976. *Images of Canadian Futures: The Role of Conservation and Renewable Energy*. Fisheries and Environment Canada, Ottawa.

Smil, V. 1974. *Energy and the Environment: A Long Range Forecasting Study*. Manitoba Geographical Studies 3, Department of Geography, The University of Manitoba, Winnipeg, Canada, 187 pp.

Smil, V. 1977. China's future: A Delphi forecast. *Futures*, **9(6)**:474–489.

Steinbrugge, K. V. 1970. Earthquake damage and structural performance in the United States. *In:* R. L. Wiegel (Ed.), *Earthquake Engineering*. Prentice-Hall, Englewood Cliffs, New Jersey, pp. 167–226.

Sutherland, S. 1975. 'Exercise London': A community disaster plan. *Emergency Planning Digest*, **2(3)**:11–12.

Taylor, J. L. Urban gaming simulation systems. *Progress in Geography*, **3**:135–171.

Thornton, B. 1971. *Gaming Techniques for City Planning: A Bibliography*. Council of Planning Librarians Exchange Bibliographies, No. 181, 14 pp.

Trudeau, P. E. 1970. Technology, the individual and the party. *In:* A. M. Linden (Ed.), *Living in the Seventies.* Peter Martin Associates, Toronto, pp. 1–7.

Turner, J. S., and D. K. Lilly. 1963. The carbonated water tornado vortex. *Journal of Atmospheric Science,* **20**:468–471.

U.S. Office of Emergency Preparedness. 1972. *Disaster Preparedness.* Report to Congress, Executive Office of the President, 184 pp.

Walford, R. 1969. *Gaming in Geography.* Longmans, London.

Whalin, R. W., D. R. Bucci, and J. N. Strange. 1969. A model study of wave run-up at San Diego, California. *Tsunamis in the Pacific.* Proceedings of the International Symposium on Tsunamis and Tsunami Research. University of Hawaii, Honolulu, October 7–10, 1969.

White, G. F., and J. E. Haas. 1975. *Assessment of Research on Natural Hazards.* Massachusetts Institute of Technology, Cambridge, Massachusetts, 487 pp.

Whitham, K., W. A. Milne, and W. E. T. Smith. 1970. The New Seismic Zoning Map for Canada (1970 ed.). *The Canadian Underwriter,* pp. 1–9.

Whitman, R. V. 1973. Damage Probability Matrices for Prototype Buildings. *Seismic Design Decision Analysis,* Report No. 8. Massachusetts Institute of Technology, Cambridge, Massachusetts.

Whitman, R. V., J. M. Biggs, J. Brennan, III, C. A. Cornell, R. de Neufville, and E. H. Vanmarcke. 1973. Summary of methodology and pilot application. *Seismic Design Decision Analysis,* Report No. 9., Massachusetts Institute of Technology, Cambridge, Massachusetts.

Wood, C. J. B. 1973. *Handbook of Geographical Games. Western Geographical Series,* 7, University of Victoria, Victoria, B.C., 154 pp.

Wuorinen, V. 1976. Seismic microzonation of Victoria: A social response to risk. *In:* H. D. Foster (Ed.), *Victoria Physical Environment and Development, Western Geographical Series, 12,* University of Victoria, Victoria, B.C., pp. 185–219.

6
Disaster Warning Systems

In nature there are neither rewards nor punishments—there are consequences.

Lectures and Essays, 3rd Series
R. G. Ingersoll (1833–1879)

The Nature of Warning Systems

The alarm clock is one of the commonest of all warning systems. Despite its relative simplicity, however, it illustrates most of the sixteen basic principles involved in the operation of very much more complex networks. As in the case of any hazard, danger must first be recognized. That individuals often overslept was well known. The alarm clock was, therefore, invented to reduce the future probability of such mishaps. This mechanism was manufactured, tested, and widely advertised. Since there was general agreement that the danger of oversleeping existed, such clocks sold well.

The mere ownership of the mechanism, however, was not enough to guarantee an early awakening. When oversleeping seemed likely, the alarm clock had to be set. Once this step had been taken, if the danger of morning lethargy became imminent, a bell or buzzer sounded to awaken the sleeper in sufficient time to take the necessary preventative action, that is, to get up. In newer models, this noise persisted until an appropriate response, the pushing of a button or pulling of a plug, indicated that it had in fact awakened the sleeper.

From this simple example it is apparent that providing warning is a complex process. It usually involves the interaction of physical, technological, and social systems, the operation of which must be carefully coordinated if the desired result, the avoidance of disaster or reduction of the scale of devastation, is to be achieved. With the exception of specialized systems, restricted in their use to military bases, power stations, industrial complexes, or similar establishments, the successful operation of a warning system almost always involves active regional and local government participation. This generalization is true, regardless of the size of the network involved. The Pacific-wide Seismic Sea Wave Warning System, for example, is international in operation yet, in the final analysis its success depends on rational response to tsunami information provided at the local level (Foster and Wuorinen, 1976). In the same manner, national, state, and provincially operated warning systems, such as the flood-related disaster networks of the United States National Weather Service, are also very dependent on local government cooperation for their success because, ultimately, the public must respond to warnings, and the police, fire, and social services capable of influencing them are generally under local government control (Owen, 1977).

Even where warning systems are privately owned and operated, as for example are smoke detectors, regional and local government initiative can be important. The city of Welland, Ontario, has recently requested that Canadian insurance companies provide reduced rates to homeowners who install such sensors, while the province in which this settlement is located now requires that all homes built after 1 January 1976 must have smoke detectors (Shaw, 1978).

A valid warning must consist of two discrete components, one of which highlights the existence of danger while the other elucidates alternative courses of action which can prevent, avoid, or minimize such risk. Rather than viewing warning as simply the linear transmission of a message, it is perhaps more realistic to accept McLuckie's (1970) view of it as the operation of a system. Morphologically, many such networks consist of observers who detect danger and predict its temporal and spatial dimensions: communicators, such as police or civil defense organizations, and the public who themselves, as message recipients, form the final vital link. Warning then is a process that begins long before a threat is detected and is completed when appropriate action has been taken by those at risk. Because all decision-making systems are learning cycles, it is essential that the adequacy of such information flows be continually reevaluated with reference to recent experience and that, where necessary, modifications be made to ensure more efficient operation. However, refinement after adverse experience has greater legitimacy in more mundane decision-making procedures, since clearly any malfunctioning of a disaster warning network is socially and financially expensive, and must be avoided, if at all possible.

Although warning systems vary in size, scope and effectiveness, reflecting their social setting and the hazard being monitored, they increasingly involve the use of sophisticated technology and the cooperation of a variety of organizations and individuals. This complexity is illustrated in Figure 6.1, which shows that the successful operation of a warning system typically involves sixteen basic steps. For the sake of convenience, these are outlined in abbreviated form on the next page and then described in detail.

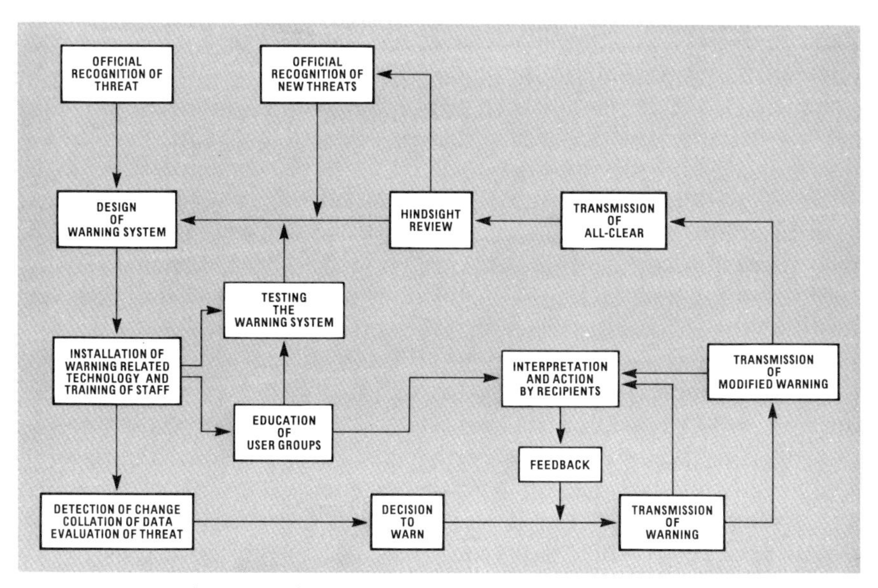

Figure 6.1. Idealized warning system.

1. Recognition by decision makers that there is the possibility of danger from a particular source.
2. Design of a system to monitor changes in the hazard and issue warnings if danger increases beyond certain thresholds.
3. Installation and operation of the system.
4. Education of the user group, often the general public, so that should a warning be issued, responses will be appropriate. The infrastructure may also have to be modified to permit effective operation.
5. Testing the system, when there is little danger, to ensure that it is technically sound and that those involved in issuing and receiving its warnings act as required.
6. Modifying the system if test results indicate that changes are necessary.
7. Detection and measurement of changes in the hazard which could result in increases in death, injury and/or property damage.
8. Collation and evaluation of incoming information.
9. Decisions as to who should be warned, about what damage, and in what way.
10. Transmission of a warning message, or messages, to those whom it has been decided to warn.
11. Interpretation of the warning messages and action by the recipients.
12. Feedback of information about the actions of message recipients to the issuers of the warning messages.
13. Transmission of further warning messages, corrected in terms of the user groups' responses to the first and subsequent messages and noting any secondary threats.

14. Transmission of an all-clear when danger has passed.
15. Hindsight review of the operation of the warning system during potential disaster situations and the implementation of any necessary improvements.
16. Testing and operation of the revised system.

As with any system, the input may not necessarily produce the expected or desired output. An increase in danger detected by the warning network, for example, does not automatically lead to the target group response required to minimize losses. Failure to respond to warning may occur for a variety of technical or social reasons, which can arise during any one of the sixteen steps outlined above.

Recognition of Danger

On 11 May 1953, Waco, Texas was struck by a tornado that caused 114 fatalities and over 1000 injuries, some 145 of them serious (Moore, 1958). The great majority of these casualties could have been avoided if as little as five minutes advance notice had been given of the approaching threat. However, despite the fact that there had been storm and tornado warnings throughout the day for the general area, little attention had been paid to them in Waco, which had no effective local response system. (Moore and Crawford, 1955). Vigilance was not considered necessary since most

PLATE 10. Tornado approaching Wichita Falls, Texas, April 3, 1964 (American Red Cross photograph).

residents believed an Indian legend, widely publicized by the Chamber of Commerce two years previously, that the city was immune to tornadoes (Dacy and Kunreuther, 1969).

In the early hours of 28 March 1964 a seismic sea wave, generated in the Gulf of Alaska, surged up the Alberni Inlet on Vancouver Island. No warning had been given to the inhabitants of the twin communities of Alberni and Port Alberni. Fortunately the first wave served to alert residents, giving them time to evacuate. Later, higher seismic waves combined to destroy 58 buildings and damage a further 320 properties (White, 1966). The lack of warning was typical of the whole west coast of Canada where losses from this 1964 tsunami exceeded $10 million. In contrast, in most places elsewhere in North America and in many other Pacific-rim countries, coastal inhabitants had received several hours notification of the approaching tsunami. That Canadians had been given no warning was the result of the country's 25 June 1963 withdrawal from the Pacific-wide Seismic Sea Wave Warning System. On that date, the Canadian Department of Transport discontinued its involvement in the network, a decision based on recommendations of the Canadian Committee on Oceanography (Erb and Wall, 1972). It had been decided, in error, that the west coast of Canada was not at risk from tsunamis. The two preceding case studies illustrate a very obvious but critical truism, that the design and installation of any warning system must be preceded by the recognition of danger.

As has been stressed previously, a major role of any safety committee is to identify all hazards that threaten its community, rank them in order of significance, and determine when and where they are likely to disrupt normal activities. It is on the basis of such information that the necessary warning systems can be established at the local level.

Designing the Warning System

In the event of a nuclear attack on Canada, its inhabitants have been advised to "duck, hide, hope and pray" (Stirton, 1971). They are to take these actions only after sirens have sounded and every radio and television station in the country has broadcast the Attack Warning. How effective this advice is likely to be is debatable, but it does illustrate that like chains, warning systems are only as strong as their weakest links. For this reason, such networks must be designed with great care. Attention should be paid to both technical and social components, their interaction and to the networks' roles in the social system of which they are merely small parts.

Well designed warning systems save lives. As a result, within the last century, in the industrialized world, the character of threat from natural hazards has changed largely from one of significant loss of life to the expectation of property damage (Dacy and Kunreuther, 1969). In the United States, for example, between 1930 and 1935 nearly three lives were lost for every million dollars of natural hazard destruction. Today approximately $4 million of property damage is associated with each life lost. As can be seen from Table 6.1, fatalities from hurricanes, floods, and tornadoes have decreased in the United States during the past century, despite increases in population. In the period 1925 to 1949, 11,234 people lost their lives from these hazards. This total fell to 7201 in the period 1950 to 1974.

Improvements in warning systems can be credited with much of this increase in safety. In 1961, for example, when Hurricane Carla was still 1150 miles from the United States mainland, emergency preparations had already begun. As a result of warnings issued, an estimated 350,000 people left the coastal regions of Louisiana and Texas and many high risk areas were almost completely evacuated. The ensuing loss of life was small, only 46 killed, despite property damage of some $400 million (Dunn and Miller, 1964). These figures stand in stark contrast to those caused by the impact of the hurricane that struck Galveston Island in 1900, leaving 6000 dead but causing only $30 million of destruction (Dacy and Kunreuther, 1969).

Where large losses of life still occur in the Developed World, these usually take place because of the lack or malfunction of local warning systems. On 31 July 1976, for example, the Big Thompson Canyon was ravaged by a flash flood after its upper and central watersheds received intensive rainfall. In some areas over 12 inches fell in less than 5 hours (Gruntfest, 1977). In the resulting flood, 139 people died, 88 were injured, and six are still missing. There was no effective flash-flood warning system in the area and most of the people in the Big Thompson Canyon were

Table 6.1. Loss of Life in the United States Due to Hurricanes, Floods, and Tornadoes, 1925–1977

Years	Hurricanes	Floods	Tornadoes	Total
1925–1929	2114	579	1944	4637
1930–1934	80	146	1018	1244
1935–1939	1026	783	921	2730
1940–1944	149	315	835	1299
1945–1949	67	304	953	1324
1950–1954	217	293	885	1395
1955–1959	660	498	523	1681
1960–1964	175	242	230	647
1965–1969	412	512	705	1629
1970–1974	146	1000	703	1849
1975–1977	64	512	200	776

Sources: Schwartz (1979); Dacy and Kunreuther (1969).

not officially informed of the danger, leading to the worst disaster in Colorado's history.

The complexity and variety of hazards and their impacts lead to great diversity in the size and scale of existing and additional necessary warning systems. There are at least nine disaster agent characteristics which have implications for warning network design (McLuckie, 1970): frequency, physical consequences, speed of onset, length of possible forewarning, duration, scope of impact, destructive potential, gross predictability, and gross controllability. Each must be considered by the safety committee when designing local warning networks.

Frequency

The frequency with which any disaster agent strikes varies greatly from place to place. For this reason an initial step in assessing the need for a local warning system must be the perusal of disaster incidence records and maps illustrating the international, national, and regional distribution of known hazards. Such information is usually produced by either official sources, such as environmental departments, geological surveys, meteorological offices, or by academics, notably seismologists, geologists, geomorphologists, and engineers. Examples include the maps of tornado frequency, seismic threat, and flash flood risk in the United States, collected and published by the U.S. Office of Emergency Preparedness (1972). Such sources may also be able to provide additional information on strictly local risks.

Familiarity with a particular hazard leads to an expertise in recognizing mounting danger. It also increases the possibility that, where no warning system exists, this deficiency will be remedied. The response of a community to any disaster agent also tends to improve with experience. A hurricane-prone city like New Orleans is especially attuned to potential threat. It has a comprehensive organizational structure to collect and collate information, disseminate warning messages, and evoke appropriate public responses (McLuckie, 1970).

Unfortunately many of the worst disasters are suffered by communities that have traditionally been peripheral to hazard-prone areas. This is because the social organization needed to respond to rare threats from vague disaster agents is absent. Safety committees, therefore, should be extremely careful before dismissing as negligible risk from hazards that have caused serious losses in neighboring states or municipalities but not in their own. In such locales, it may also be difficult to generate public interest even when warnings have been issued. Darwin, Australia, had often received tropical cyclone alerts or warnings. Normal storm paths, however, saved the city from disaster on many occasions and little cyclone damage had been suffered since 1937. As a result, when cyclone warnings were issued on Christmas Eve 1974, most residents believed that, as usual, the city would be spared and so continued their normal festivities (Haas, Cochrane, and Eddy, 1976). This proved to be an expensive error since the city, on this occasion, suffered extensive hurricane damage which led to its eventual evacuation.

It is difficult for those in charge of disaster warning systems to deal with the complacency generated by several near misses. One strategy is for the safety committee to release detailed explanations of why disaster did not occur after a warning had been issued and to provide detailed descriptions of what would have happened if the area had been less fortunate.

Consequences

Hazards may be geophysical, meteorological, biological, or man-made and cause destruction or injury in an enormous variety of ways. Such physical differences and the consequences of disaster agents naturally have considerable implications for the design and operation of warning systems. Nevertheless, it is essential that the network be able to function before, during, and after disaster strikes. In the past this has frequently not been the case. The 28 March 1964 Alaskan earthquake, to give an example, destroyed the control tower at Anchorage International Airport, breaking vital communication links between the nerve center of the Seismic Sea Wave Warning System in Honolulu, the College and Sitka seismic observatories, and the Kodiak, Sitka, and Unalaska tidal stations. As a result the receipt of data for these locations was delayed and the functioning of the system was impeded (Spaeth and Berkman, 1967).

Since many initial disasters trigger further catastrophes, it is essential that warning systems are designed to function after primary impact. The earthquake that devastated the cities of Yokohama and Tokyo on 1 September 1923 started numerous fires in both cities. Communications were destroyed and authorities were unable to prevent refugees congregating in higher, less damaged parts of Tokyo. More than 40,000 people crowded into the Military Clothing Depot, an open space where all perished, suffocated by the searing air from which the oxygen had been withdrawn by earthquake-induced fires (Hodgson, 1964). It is therefore imperative that warnings detail exactly what physical consequences may be expected. This is especially true in the case of chemical hazards, industrial fires, damage to nuclear plants, or other disaster agents with which the public is unfamiliar.

Speed of Onset and Length of Forewarning

Three types of hazard can be differentiated on the basis of their speed of onset. This variable, in turn, influences the length of forewarning possible. In the case of rapid onset, the length of time between the preimpact

PLATE 11. Extensive flooding at Tonbridge, Kent, after 36 hours of torrential rain. Warning systems must be established in flood-prone areas. They must be capable of functioning after primary impact (British Information Services, Central Office of Information photograph).

phase and the emergency period is very short because the agent strikes swiftly. In some cases, such as earthquakes, there has traditionally been no time in which to issue a warning. It should be noted that the length of forewarning about the onset of a hazard is a function of knowledge. As the theory of seismology improves, it is becoming increasingly feasible to predict earthquakes perhaps years ahead of their occurrence (Working Group on Earthquake Hazards Reduction, 1978). Similar advances in the use of Doppler radar may soon lengthen the forewarning period of tornadoes by over half an hour (Burgess, Hennington, Doviak and Ray, 1976).

Gradual onset occurs when the effect of a hazard slowly increases in intensity until an emergency period is reached. To give an example, droughts may take several months, perhaps years, to reach this stage. Spring floods, caused by the melting of heavy winter precipitation, can often be forecast months ahead by monitoring the depth of the snow pack. Forest fire hazards also increase in a similar cumulative manner as the timber and other fuel becomes dried. A third type of disaster agent has a repetitive onset, striking several times in rapid succession. Tsunamis are typically of this type. Figure 6.2 shows the length of forewarning in Honolulu for seismic sea waves generated around the Pacific Ocean.

Both the probable speed of onset and the length of forewarning are dimensions that must be considered by safety committees in the design of disaster warning systems. When the hazard strikes rapidly and the warning period is short, technology must function in real time. Fewer people are likely to receive the message of impending danger, and their reaction must be almost instantaneous. Many adjustments to the threat, such as evacuation before the hazard strikes, are simply impossible.

One interesting innovation is beginning to improve disaster response to hazards of rapid onset. In the United States the National Oceanic and Atmospheric Administration has designed and is now operating a VHF-FM Tone Alert Warning System. This ties into the regular weather warning activities of the National Weather Service. A government operated radio transmitter will, when activated in response to predicted disaster, broadcast a coded signal. This will be picked up in the broadcast area by home receivers, bought by residents from private industry. This signal will trigger a spoken description of the anticipated hazard which will be preceded by a siren sound issuing from the set (Lewis and Clark Law School, 1977). Similar systems are being studied by the Defense Civil Preparedness Administration and by the National Aeronautics and Space Administration. The latter research involves the use of satellites to monitor hazards and trigger home receivers. In communities subject to hazards that can strike with little forewarning, the safety committee should encourage the purchase of such equipment when it becomes available. Special emphasis might be placed on ensuring that it is available in public buildings, such as libraries, schools, hospitals, and government offices.

It should be remembered that a slow speed of onset does not necessarily lead to effective warning. Too long a period of forewarning, unaccompanied by any obvious signs of mounting danger, may result in an apathetic public reaction since the hazard-related information is not taken seriously. In extreme cases, as with the several years of forewarning that may eventually accompany earthquake predictions, the preimpact time period may be accompanied by large forecast-related economic losses, proving to be a very mixed blessing (Working Group on Earthquake Hazards Reduction, 1978).

Duration

The impact of disaster agents may vary from virtually instantaneous to protracted. Explosions are an example of the former and forest fires of the latter. The type of information provided to the public by a disaster warning system must vary with the duration of the damaging event. This is because the precautions that must be taken are influenced by the duration. As McLuckie (1970) has pointed out, some of the difficulties that oc-

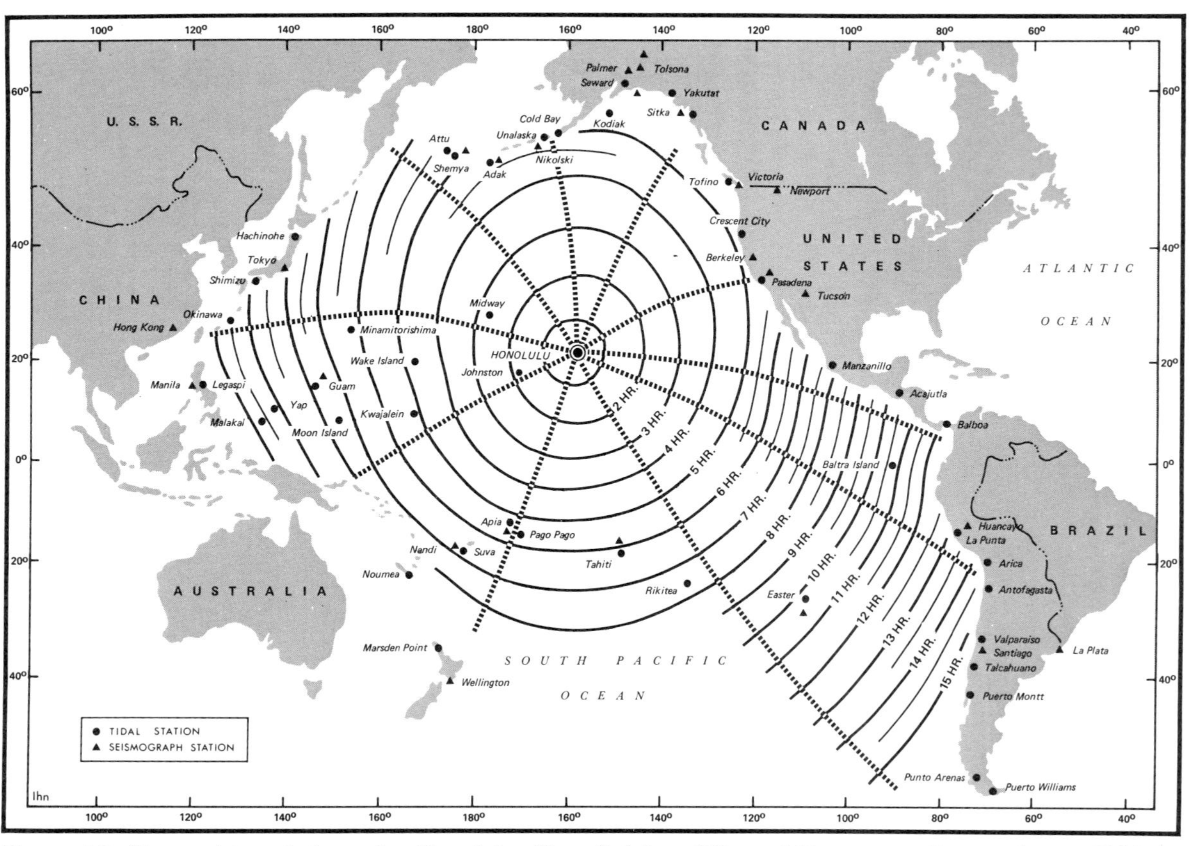

Figure 6.2. Tsunami travel times for Honolulu, Hawaii (after Office of Emergency Preparedness, 1972, a predecessor of the U.S. Federal Emergency Management Agency, *Disaster Preparedness*).

curred in New Orleans as a result of Hurricane Betsy resulted from the fact that people were only warned about, and prepared for, the initial storm. They were not ready for the much longer duration flood associated with it. Since communications tend to fail as the result of the impact of an initial disaster agent, it is also often difficult to transmit effective secondary warnings during emergencies of long duration.

Scope of Impact

The area threatened by a hazard may be very localized or extend over thousands of square miles. In examples of the former it is often possible for local warning systems to be supplemented with outside aid. This was the case on 14 December 1963 when a break in the Baldwin Hills Dam in Los Angeles devastated one square mile of residential and business properties (Anderson, 1964). The limited area threatened allowed police officers and personnel from the fire department to go from door to door evacuating residents in the most critical sections. The police department alone utilized 512 men, drawn from all divisions in Los Angeles. In many localized disasters, communication equipment and manpower might remain almost intact and be capable of issuing secondary warnings.

In contrast, diffuse impacts may render systems isolated and incapable of generating further warnings. The Easter Sunday tornadoes that cut across Illinois, Indiana, Michigan, and Ohio affected such a large area that it was difficult for earlier devastated communities in the west to warn eastern localities still at risk (McLuckie, 1970).

It is, therefore, extremely important for communities to attempt to predict the scale of damage particular disaster agents might be expected to cause. This should include estimates of destruction in the surrounding region. Such information then permits a realistic assessment of both the probable availability of outside help and the likelihood of requests for aid that may be received from elsewhere.

On the basis of this assessment of damage potential, mutual assistance agreements can be negotiated with communities that would probably be outside the area of impact of potential local hazards. This is necessary since, in times of disaster, prior commitments made to and with adjacent municipalities or districts may be impossible to honor because of widespread damage.

Destructive Potential

Disaster agents vary dramatically in their destructive potentials. Some, such as the landslide which destroyed much of the town of Frank, Alberta, can remove all traces of life, property, and even the original ground surface. In contrast epidemics, poisons, gases, or radiation may be responsible for large losses of life but little, if any, property damage. The reverse may also occur. The spread of exotic species, hailstones, or frosts can cause extensive agricultural losses without threat to human health.

These differences in destructive potential are important in the design of a warning system. Prior to impact, it is not always easy to gauge the full range of threats associated with a disaster agent and thus to evaluate the kind of warning that should be issued. Many factors may affect the potential destructiveness of a tornado funnel or hurricane cloud, for example, even if they do hit a particular locality (McLuckie, 1970). Because the precise nature of some hazards cannot be predetermined, the secondary threats that may be associated with them, such as the lines of tornadoes frequently spawned by hurricanes, are all but impossible to predict before impact. Where the operators of a warning system are unsure of the destructive potential of a hazard they would probably be well advised to run the risk of overestimating its consequences.

Gross Predictability

The ability to predict the occurrence of disaster agents varies enormously. Most explosions, for example, are outside any available forecasting methodology. Others such as tornadoes, forest fires, and floods can be partially anticipated. In these cases, gross predictions can be made of the probability of occurrence, general paths likely to be taken, and possible intensities of impact. More specific details cannot yet be forecast, although research continues to improve understanding.

Gross predictability of the threat from any disaster agent greatly influences the kind and length of warning that is possible. This, in turn, influences the whole process of response. From the point of view of local government, what should be stressed is that gross predictability is continuing to improve and every effort must be made by the safety committee to monitor advances in theory and technology so that the benefits may be passed on to the community as soon as possible.

Gross Controllability

Some disaster agents can be eliminated or greatly influenced, for instance, forest fires or floods; while others such as earthquakes or hurricanes cannot. As technology improves, many hazards may eventually be brought under human control. The current degree of gross controllability also influences the warning process. If an agent is potentially controllable there may be a reluctance to warn and alarm people. This occurred in the case of the Vaiont Dam, where on 9 October 1963, a huge landslide fell into its impounded reservoir. The displaced water overflowed the dam and engulfed the town of Longarone and more than a dozen other hamlets in northeastern Italy. Some 2600 lives were lost (Kiersch, 1965). That the slide was imminent was well known but it had been erroneously believed that the reservoir could be emptied before it occurred. For this reason no general alarm had been raised.

If no control is possible, warnings are more likely to be issued or residents will respond to evidence of increasing danger on their own accord. Since most of the work on the gross controllability of hazards is occurring at the international or national levels, local responsibility rests with monitoring the literature and accepting beneficial innovation as quickly as possible.

Assistance

It is very unlikely that a safety plan coordinator and his committee will be faced with threats from hazards unknown elsewhere. For this reason experience from other areas can be of great value in designing a warning system.

To illustrate, local governments can frequently call upon higher levels of administration for assistance in setting up disaster warning systems. Many senior agencies will provide information, planning and technical aid, and even financial support for such activities. Owen (1977) lists seven federal organizations willing to assist local government in the United States to set up flash flood warning networks. These include the National Weather Service, the Federal Disaster Assistance Administration, the Comprehensive Planning Assistance Program, and the Soil Conservation Service. Other useful sources of information include research institutions, such as the Natural Hazards Research and Applications Information Center at the University of Colorado in Boulder (Torres and Waterstone, 1977). It is also helpful to contact other local governments that have suffered disaster losses and have therefore had the opportunity to test and improve their warning systems as a result.

Personnel

Since many key personnel may be ill, on holiday, or stranded when danger threatens, it is always useful to design a warning system in which responsibility passes automatically from one individual to another if absence makes this necessary. Such a chain of command is particularly necessary since many individuals directly involved in the operation of the network may be killed or injured by the initial impact of the disaster agent, so making the issuing of secondary warnings difficult. Preplanning of the hierarchy of authority therefore saves valuable time and confusion in the event of such a contingency.

Needs of Special Groups

As Riley (1971) points out:

> Hurricane Camille was far from impartial. Flood tides showed a strong preference for older people; 17 percent of the fatalities were over the age of 75 and 37 percent over 65. When compared with the total number of people in the various age brackets, the vulnerability of the older citizens becomes even more shocking.

Fatalities claimed by Hurricane Audrey again illustrated the great vulnerability of the elderly (Friedsam, 1960). Very young people are also at greater risk than the general population.

In the design of any warning system it must be remembered that the population is far from homogeneous. The deaf cannot be expected to respond to an alert system based on sirens, nor can the patient in hospital bed, or the individual who is too infirm to leave home, evacuate without special assistance. Similarly, ethnic groups may speak a different tongue from the majority and must be warned in their own language.

It is suggested that the safety committee collect data on the educational level, age structure, and health of the community. This will allow the pinpointing of high risk individuals and locations such as hospitals, nursing homes, and schools where special assistance will be needed in the event of an evacuation. It should also be remembered that in the case of certain chemical threats, pregnant women should be among the first to leave the area.

Legal Responsibilities

Through legislation, agreement, or precedent, many local governments are obligated to operate warning systems. Failure to provide a warning where a legal commitment exists may prove extremely costly in the event of later litigation. The reverse may also be true. When incorrect warnings have been given and unnecessary disruption to trade or recreation has occurred, a municipality may find itself held liable for the costs incurred. For these reasons, it is essential that legal counsel be sought before any decision is taken to establish, close down, or significantly modify any disaster warning network. In addition, the type of warning messages should be subjected to prior legal scrutiny to ensure that, in the event of litigation, a municipality is unlikely to be held negligent.

Installation and Operation

The type and distribution of sensors used to collect data are critical to the efficient functioning of any warning network. There is often a minimum number below which the system is incapable of fulfilling its obligations.

Ideally, these sensors should be spaced sufficiently close together to give a realistic picture of the buildup of threat. Where response to threat must be rapid, information should be telemetered automatically to a central processing center. Such adequate coverage rarely exists; for example, during the initial states of the Big Thompson flood, Denver staff meteorologists had considerable trouble trying to determine the size of the problem created by heavy rainfall (Gruntfest, 1977).

It must always be borne in mind that such equipment may not always function effectively due to the impact of the hazard itself, or to unrelated technical problems or sabotage. During the severe flooding in the midwestern United States in 1969, some of the worst losses were suffered by Minot, North Dakota. Many of the data necessary to predict the flood levels of the Mouse (Souris), the river responsible for damaging two schools, 15 churches, 212 businesses, and the homes of 11,864 people, were unavailable since Canadian gauging networks in the headwater of the river had been drowned out (White, 1969).

With the recent increases in vandalism and sabotage, it appears likely that many unbalanced individuals may damage or destroy sensing equipment to decrease the preparedness of threatened communities. This is most likely to be the case if the hazard itself has been deliberately created. Flood (1976) has pointed out that the twentieth century has already witnessed some 100 million deaths caused by our fellow men. He argued that the last 75 years appear to have been the most violent in history. Terrorism is being used as a lever to influence government and arson as a method of gaining wealth from insurance companies. To illustrate this point, Flood lists 11 terrorist attacks against nuclear installations or facilities that took place between 1969 and 1975 on a worldwide basis. The creation of disaster has become a political tool. Every precaution should be taken to ensure warning equipment remains functional.

Data processing and disaster warning centers should be located in very low risk zones established by microzonation. It is essential that they continue to function after the impact of a hazard. For this reason it is generally unwise to locate such facilities on flood plains, within one mile of the coast or lakeside, or on surficial sediments in seismically active areas. Related communications equipment must also be located with similar care.

Ideally, every warning network should be supported by an independent back-up system. This should be capable of operating in isolation if the primary network is rendered inoperative or incapable of functioning effectively. For example, the author (Sewell and Foster, 1975) has suggested that the tsunami warning system in British Columbia, essentially based on radio and telephone communications, should have as its back-up system aircraft flying above the speed of sound, followed by slower planes dropping flares. Such a system could alert wide areas in a very short time, if the primary means of warning malfunctioned.

Education of the User Group and Modification of the Infrastructure

In disaster situations there is no substitute for knowledge. When flash flooding devastated the Big Thompson Canyon, Colorado, on 31 July 1976 some of those threatened tried to escape by driving out of the canyon. The high velocity flood waters overtook many of these, drowning them almost instantaneously. Had they abandoned their cars and climbed 25 ft up the canyon wall they would most certainly still be alive today, little the worse for their experience (Gruntfest, 1977). Such error under stress is a common element of disaster, as could be illustrated by numerous other examples.

While there is no guarantee that those warned will take appropriate action, the likelihood of their survival can definitely be improved by predisaster education programs. Ideally, since time is generally very limited in crisis situations, those who receive a warning should be both willing and capable of instantaneously taking appropriate action. What is in fact required is the military response to authority, an unhesitating obedience to command. Such obedience is alien to most members of the public and predisaster education is needed to persuade them that it is necessary in very dangerous situations. The value of this willingness to act on command was made apparent by the collapse of the Baldwin Hills reservoir, which was not visible from built-up areas because of its sunken hilltop location (Anderson, 1964). Since many people were unaware of its existence, it was difficult to persuade them of the likelihood of flooding especially since the weather had been fine. Public education should not simply aim at developing a trust in authority, since in many disaster situations, self-reliance is essential. There is also an individual need to recognize and respond to threat clues, signs of impending disaster, such as changes in wind, noise, rapidly rising river levels, or the sudden and unusual death or illness of small animals and birds. It should be remembered, however, that the more nervous members of the public may begin to panic prior to any disaster situation, if such a program is mishandled.

Ensuring that public response to disaster warnings is optimal is no simple matter. This task is complicated by the great differences in experience, age, intelligence, health, and interests represented in the resident population. In addition, in tourist areas many of the public may be transients. Any effective education program must recognize these difficulties and try to serve as many distinct groups as is feasible.

Fortunately, numerous avenues of influence are available to the safety committee. These include newspaper and magazine articles, television and radio broadcasts (with special emphasis on open-line shows where the public can question experts), films and videotapes, telephone answering services (either live or precorded), directories (including the yellow pages), pamphlets, mobile exhibitions, guest lectures, bulletin and notice boards, and school, club, and society visits.

Perhaps the most critical point to be made is that, regardless of the medium, the message must be presented at a level that can be readily understood by those members of the public at which it is aimed. This may seem self-evident, but it is often forgotten. For example, it has been estimated that over 90% of Americans cannot understand their federal income tax forms, despite the fact that annually they are obliged to complete them.

Information must be presented to educate, not to impress. For this reason, it is advisable to employ or seek the aid of reporters who are skilled in public communication, when preparing disaster information kits, films, or articles for newspaper distribution. Similarly, children have a limited vocabulary and often appear to understand concepts that in fact they have failed to grasp. Trained teachers should assist in preparing materials aimed at this level. It is, however, still advisable to pretest the message by submitting it to a small group of the desired audience before general release. Schwartz (1979) has provided a graphic example of the value of such disaster education. In 1974, during a tornado outbreak in the United States, a seventh grade teacher reviewed severe weather safety rules with his class. After school, as one of the buses was taking pupils home, a tornado appeared in its path. The driver did not know what to do, but one of the recently briefed students did. He convinced the driver to pull over, get everyone out and far enough away so that the bus couldn't roll onto them. All passengers sought safety in a ditch. Although the tornado destroyed the bus, none of the pupils was injured.

It may not be necessary for the safety committee to prepare its own warning literature since many such pamphlets and films are already in circulation. Examples include brochures produced by the U.S. National Oceanic and Atmospheric Administration, the U.S. Geological Survey, and the Canadian Emergency Measures Organization. Particularly impressive is the literature being distributed by the Texas Hurricane Awareness Program. Such materials can either be used as models for the design of more locally focused literature or may be obtained in bulk for public distribution. Morton (1978) has produced a directory of sources of films and other visual materials on natural hazards and their mitigation which will be of value to safety committees.

Brochures, newspapers, and other similar materials have a short lifespan. It is therefore also useful for disaster warning information to be included in the yellow pages or elsewhere in the telephone directory, as is the case with tsunami response instructions in Oahu, Hawaii. This guarantees a wide distribution and survival for at least a year. Public noticeboards warning of local hazards and providing response information may also be erected. This is one of the few means of providing disaster information that is readily available to tourists, who in some areas may outnumber local residents at the height of the season. Since the collective public memory is notoriously short, disaster education programs should be conducted on a continuing basis wherever manpower permits.

One further point should be stressed: it must be absolutely certain that in the event of an emergency, the public could realistically follow the instructions provided. For example, in the city of Victoria, teachers have been urged to telephone parents immediately after a severe earthquake. Such a course of action is likely to be impossible since the city's telephone communications system is almost certain to be badly damaged in the event of such a seismic episode. Even if in the unlikely event that it were not, attempts to contact several thousand parents would tend to overload the remaining circuits, at a time when they would be critically needed for more urgent purposes. In addition, all agency disaster literature must be coordinated since conflicting official instructions can lead to chaos.

The distribution of information should be accompanied by an ongoing assessment of its impact. To achieve this objective a short questionnaire might be designed by the safety committee to discover whether disaster information is being understood, and if it appears likely to lead to the desired response in an emergency. Such a questionnaire survey might seek to establish which hazards the public perceive, how they expect to be warned when threatened, and what action they would expect to take on receiving such information.

This questionnaire might be administered at shopping centers or be included with other local government mailings. The former alternative is preferable, since questionnaires distributed by post are more likely to be returned by the more knowledgeable members of the sample, perhaps leading to an overestimate of public preparedness. Naturally, where the questionnaire reveals misunderstandings, every effort should be made to correct these through a redesign of the disaster education program. Where it is clear that the public is well informed, however, the questionnaire survey must not be used as justification for reducing the hazard information program.

Testing the System

Disaster warning networks, because they are essentially learning systems, generally improve with use. Unfortunately, since the social cost of a malfunction is so high, it is not advisable to leave their testing until disaster situations occur. Improvements should also be based on the results of disaster simulations.

Testing the technology on which a system is based does not constitute testing the network. It is, however, a critical first step. Radar, river gauges, seismographs, and all other physical plants must be maintained and, as far as possible, kept continuously operational. This is illustrated by one of the worst disasters of the twentieth century, the 12–13 November 1970 Bangladesh cyclone. The peak of the accompanying storm surge nearly corresponded to high tide and the winds and associated flooding claimed at least 225,000 human lives, caused $63 million damage to crops and killed 280,000 head of cattle (Frank and Hussain, 1971). Although the cyclone had been identified by neighboring meteorological services as early as 9 November and tracked by satellite and later by radar as it moved northeastward up the Bay of Bengal, the initial warning was not passed on by the local radio station which closed at 11 P.M. Moreover, a newly adopted streamlined system of warning was resented by officials, high and low, who blocked it (Burton, Kates, and White, 1978).

The operation of warning-related technology should be tested at regular intervals, as should back-up systems. The international tsunami warning system, for instance, is regularly supplied with "dummy messages" to determine transmission time under various conditions (U.S. Department of Commerce, 1971). In addition to the testing of a warning network's hardware, the performance of its personnel must also be monitored with a view to increased efficiency. Role play gaming and field exercises can be used for this purpose.

Modification after Testing

Disaster gaming and field exercises may dramatize several types of weaknesses in any warning system. These most commonly stem from an inadequate data collection and processing technology base, staff shortages, and interpersonal or interagency rivalry.

Shortages of staff and equipment are generally a reflection of inadequate budgetary support. However, many sources of funding exist outside the more obvious channels. Certain federal, state, or provincial agencies will help fund local warning networks or supply personnel to assist in their operation. Where such support cannot be obtained, many societies and clubs, particularly those noted for public service, might be approached to raise funds for equipment or to supply volunteers. The Susquehanna River Basin Commission's (1976a) planning guide for a self-help flood forecasting and warning program for the Swatara Creek watershed in Pennsylvannia, for example, includes the suggested use of volunteer stream gauge and rainfall observers. Similarly, the *Neighborhood Flash Flood Warning Program Manual*, prepared by the same commission, includes volunteer neighborhood coordinators, telephone call-up systems manned by those at risk, and street captains to undertake house-to-house warnings (Susquehanna River Basin Commission, 1976b). Most of the equipment necessary to monitor threat also has research potential, or can be used in the daily routine of other agencies. Such multiple use of equipment broadens the possible base of financial support to include other levels or departments of government, universities, and foundations. Safety plan coordinators may also be able to negotiate other cooperative arrangements to reduce further weaknesses in warning networks.

Detection of Threat

During all but the very recent past, the individual has been largely responsible for his own safety. This reliance on personal observation supported by the "wisdom" of legends and folktales has been termed by McLuckie (1970) the Paul Revere approach to warning. This self-reliance has been very largely superseded, in the Developed World, by government responsibility for public safety. Much of the power for the detection of threat now rests with the operators of sophisticated technology, such as satellites, medical equipment, radar, and computers, which are beyond the control and often the understanding of the general public.

This reliance upon international, national, and regional governments to detect threat has both advantages and disadvantages. Obviously highly trained personnel with access to the latest technology are generally more able to predict potential disasters and to initiate the necessary precautions than are isolated individuals, without the time or resources to constantly scientifically monitor the environment.

Conversely, however, this reliance on government can lead to a public unwillingness to accept personal responsibility for safety in the belief that, regardless of the ultimate consequences of their actions, they will be rescued by "the authorities." Such a bureaucratic approach to the detection of threat contrasts markedly with that in the People's Republic of China (Coe, 1971). Here, a large earthquake prediction program is in operation which, in addition to scientific involvement, also includes the cooperation of thousands of peasants, who monitor water levels in wells and keep track of anomalies in animal behavior. Parallels can be drawn with some warning networks in the United States where tornadoes are reported by volunteer spotters (U.S. Office of Emergency Preparedness, 1972). More public involvement in other systems would lead to a better awareness of hazards and greater support for necessary countermeasures. Public participation in local warning networks might also help to reduce the growing influence of irrational means of disaster prediction, such as horoscopes, biorhythm charts, Tarot cards, tealeaf readings, and the pronouncements of mystics.

The recognition of danger can take place at any level of organization, from the international to the local. From the point of view of local authorities, this complexity leads to problems of coordination. It is therefore necessary after all the potential hazards in the region have been identified that the safety plan coordinator contact the senior agencies involved in relevant threat detection activities, with a view to gaining access to their ongoing research and identification programs. Formal and informal communication channels should also be established. Such sources may also be invaluable in providing funding and/or expertise in setting up locally controlled warning networks.

Another characteristic of threat detection is that the groups and agencies involved in monitoring the environment often do not primarily do so for the purpose of hazard identification (McLuckie, 1970). This can lead to warning signs being ignored or down-played since they are detected by individuals involved with pure research, product design, or some other problem. For this reason, it is useful to include representatives of such groups on the safety committee.

One of the major yet, in a rapidly changing world, perhaps inevitable weaknesses of the process of threat detection is that not all hazards are being monitored. This lack of comprehensiveness in threat detection often stems from a failure to recognize new and growing dangers. One of the worst of these involves the large scale and relatively unrestricted movement of extremely dangerous chemicals along many road and rail networks. This problem has been discussed in detail elsewhere in this volume.

Threat detection is often complicated by two opposing requirements: the need for speed and the need for accuracy. For this reason it is useful to establish, well in advance of any possible disaster, just what constitutes sufficent evidence of increased threat to issue warnings. This may prevent disruptive disagreement during periods when the potential for disaster is unclear. It also requires a locally based data bank of past disaster experiences. Once such standards have been established they should be submitted to the appropriate political and agencies decision-making bodies for ratification.

Collation and Evaluation

Although most disaster-related issues have dimensions ranging from the geophysical and chemical to the political and psychological, there are few organizations at even the international or national levels that act as central locations for the collation of incoming data about threat clues from the total environment. Weather services collect data on hurricanes and tornadoes, water resources branches or departments on flooding and pollutants, forest services on brush fires, and hospitals and medical centers on epidemics. There is generally no one major data monitoring organization with a mandate to collect and collate incoming danger signs, so that an overview of total community safety can be obtained. This can inevitably lead to failure to recognize the interrelatedness of some threats. It is this role that the local safety committee should seek to play.

The collection and collation of data is a means, rather than an end in itself. However satisfactorily this process is carried out, it becomes meaningless if the information's significance is not evaluated by qualified individuals. The major problems to be faced at this stage in the operation of the warning system are those of assessing the reliability of the information; estimating the precise implications of the data; deciding if the danger indicated is such that a warning should be issued and, if so, to whom, how, and when. Any contradictions between data sources must also be resolved. In short, those responsible for operating the warning systems must decide if sources are reliable and what danger, if any, their data implies to certain places as specific times.

Evaluation therefore is not a simple process. It requires value judgment based on experience. In many cases the problems involved can be simplified by the use of models, either physical or mathematical, that allow simulations of the significance of incoming information. Examples include the use of the unit hydrograph to predict flood stages, or the construction of hardware models of actual landforms to simulate natural conditions. Such modeling has already been discussed.

A common problem in predisaster situations is an inconsistency among incoming data. There is a well-established psychological principle that, when an individual is faced with conflicting statements, he or she is likely to accept as more valid those which are least threatening (McDavid and Harari, 1968). The same is thought to be true of organizations. For this reason, evaluating groups faced with data inconsistencies tend to underestimate threats. An example of this reaction to uncertainty occurred during the early phases of the 1972 Wyoming Valley Flood, caused by Hurricane Agnes (Mussari, 1974). When the flood forecasters of the U.S. National Weather Service predicted a record-breaking 40-ft crest a few hours away from inundating Luzerne County, local officials refused to accept the prediction. Such a flood could have been expected to overflow every dike in the area. They therefore recomputed their available data and predicted a 38-ft crest. Civil defense officers used this more acceptable level to prepare for flooding in Wilkes Barre, assuming that the federal information was in error. Enormous destruction followed this decision (Cullen, 1978). Similarly, when the Japan Meteorological Agency was informed of the generation of a tsunami off the coast of Chile in 1960, it ignored the warning from Hawaii and failed to urge any local precautions. This decision was based on the fact that a 1933 tsunami initiated off Japan produced a wave amplitude of only 1 ft in Chile. The analogy did not hold true and wave height from the 1960 tsunami reached almost 20 ft in Japan. In total, 123 people were killed, 974 injured and over 5000 homes destroyed (Tsuruta, 1963).

Since the psychological principle that agencies tend to underestimate threat when data is inconsistent seems well established, it would appear prudent to always accept the most pessimistic information as correct. If this eventually proves to have been an error, unnecessary disruption and discomfort is likely to be the most unfortunate ramification. If not, hundreds and perhaps thousands of deaths and injuries may have been averted.

The Decision to Warn

The prediction is a forecast that an event of specific magnitude will take place at a certain time and location. It does not necessarily imply that any steps should be taken as a result of this information. A warning, on the other hand, is a recommendation or an order, based on such a prediction, to take precautionary, protective, or defensive action. The decision to warn therefore carries with it a great deal of responsibility. Once any organization has issued such a public pronouncement, particularly if it is based upon the prediction of an event of great destructive potential, that agency and the public's response to it will never be the same again. This generalization holds true whether or not the warning proves correct. For this reason, the decision to warn cannot be taken lightly and is generally associated with great institutional tension.

In the face of potential danger, four general scenarios can be envisaged. In the first, although some apparent threat clues are collated, no decision to warn is taken. If disaster does not follow, the information is unlikely to be made public and the agency or organization will continue to operate as before, perhaps in a more conservative manner. Conversely, disaster may in fact strike, as was the case on 9 October 1963 when the Vaiont Dam was overtopped (Kiersch, 1965). In such circumstances injuries and property losses will generally be attributed to the negligence of the officials involved. Loss of employment, status, and potential support is likely to follow and legal action may be taken by victims or their relatives (Lewis and Clark Law School, 1977).

A third scenario takes place when the decision to warn is made but the envisaged disaster does not follow. Such an error will probably result in unnecessary anxiety, inconvenience, and loss of time and money by those warned. Since hindsight will reveal that this was unnecessary, those responsible may be subject to bitter criticism and even legal action. Such mistakes undermine the future effectiveness of the warning system in two ways. Those in charge may become too cautious and fail to respond to future threat clues. Conversely, people who have wasted time and money as the result of an earlier false alarm are probably less likely to respond to future warning messages. This is particularly true if the area has never suffered devastation from the predicted disaster agent.

A fourth scenario can be constructed in which warning is given and the anticipated disaster agent subsequently strikes. Under these circumstances the adverse side effects are likely to be ignored and the value of the warning network confirmed. This point is illustrated by an amusing story recounted by Roberts (1972). According to this author, a lady called up the head of the Kansas City tornado forecasting center to complain that she had spent five hours in her basement unnecessarily as the result of a tornado watch that they had issued. Five months later, very angry, she rang again when a warning had been given but no tornado damage suffered. Two years later, after a really major storm, the same lady emerged from her shelter to find that her house had blown away. She then rang the Kansas City forecasting center once more to tell its chief, "Now, that's more like it!"

It is perhaps surprising that despite the critical nature of the decision to warn there is no widely accepted procedure upon which this can be based. What follows is an attempt, by the author, to fill this void. Although the example provided involves deciding whether or not to issue a public earthquake warning, the methodology used has as much validity for any other hazard or location.

The concept of acceptable risk has already been discussed. Life is not sacrosanct, and it is possible to calculate the number of lives lost for every million tons of coal mined or million passenger miles traveled by land, sea, or air (Table 2.1). All societies must be willing to pay a certain price in terms of human suffering for the economic and other benefits received from such activities. It is only when anticipated life loss and injury reaches unacceptable limits that any society is willing to disrupt its normal patterns of activity by issuing warnings. The following methodology is designed to establish, in a logical manner, when those limits are likely to be exceeded.

Any disaster warning system must prevent more loss than it causes. This truism, however, presupposes that the impact of an unanticipated calamity can be directly compared with one for which prior notice was available. In addition, the losses due to an erroneous warning must also be assessed. In the case of many hazards, comparisons are made more difficult by the disparate elements involved, including deaths, injuries, disruption of life style, and structural damage (Foster, 1976).

Earlier in this volume a detailed computer simulation of earthquake damage was presented for the City of Victoria (Foster and Carey, 1976). By applying the model developed by Whitman to the predicted patterns of damage it is possible to forecast the number of fatalities and injuries that might be expected from seismic events of differing intensity (Table 5.3.). Although the casualty figures presented here are based on assumptions concerning probable building occupancy rates, and can only be considered very rough estimates, they do provide some indication of the possible social impact of, for example, Victoria's 100-year earthquake. It appears likely that this event will be associated with perhaps two fatalities and 148 injuries among the resident population. A more severe earthquake, reaching modified Mercalli intensity VIII in most susceptible zones, might cause 41 fatalities and 590 injuries. In the unlikely event of intensities reaching IX in these areas, the casualty figure would rise to approximately 946 deaths and 4260 injuries.

Using the formula presented in Chapter 2, it can be established that the total stress directly caused by the 100-year earthquake in the city of Victoria is likely to be approximately 3,950,000 units, an event very comparable in size to the disaster caused by the Xenia, Ohio, tornado. If modified Mercalli intensities were to reach VIII in the most vulnerable microzones, some 4,980,000 stress units might be anticipated, resulting in a disaster comparable in magnitude to the Rapid City floods. Neither of these estimates includes the stress that might occur from secondary disasters, such as fires or landsliding.

Prediction should result in appropriate action. In the case of an earthquake forecast, a minimum response might involve drawing down reservoirs. The contents of oil and chemical storage tanks should be emptied and transported from the area. Structurally suspect buildings might also be buttressed and emergency services put on alert. If such a course of action were undertaken as the result of an earthquake warning issued for the city of Victoria, it would cause some minor inconvenience and life style disruption, perhaps equivalent to a per capita infrastructural stress value of 5. It would be extremely valuable to establish this value more precisely in areas where such precautions are being taken. This minimal response to an earthquake warning would cause some 300,000 units of stress in Victoria.

If, on the basis of anticipated damage and casualty figures, a more dramatic reaction seemed appropriate, partial evacuation might be undertaken. In the case of Victoria, this would logically include the closing of all buildings shown by the appropriate earthquake simulation to be likely to suffer more than moderate damage (Figures 5.3. and 5.4). Such a response to an earthquake warning would cause considerable disruption of the economic and social life of the city. It is estimated that it would be associated with a per capita stress of some 17, or approximately 1,020,000 units in Victoria's population as a whole. In the event that complete evacuation was deemed necessary by authorities, per capita stress might be expected to rise to 30, totaling some 1,800,000 units for the area. Such per capita estimates, of course, also require field checking when the opportunity arises.

Not only would such responses cause stress, they would also prevent it, if the predicted earthquake occurred. In the case of Victoria, by buttressing buildings, lowering reservoir level, emptying storage tanks, and taking other preventative measures, damage from the 100-year earthquake might be reduced by some 10%, or perhaps 900,000 units of stress. Evacuation of insecure buildings would also prevent most injuries, or approximately a further 40,000 stress units.

From these calculations, it can be seen that the appropriate response to a prediction of the 100-year earthquake in Victoria is the minimum discussed, that is, precautionary structural measures. Such a course of action may cause some 300,000 units of stress and prevent 900,000, a benefit–cost ratio of 3:1. Partial evacuation would create 1,020,000 units of stress and prevent only 940,000, while total evacuation would be obviously inappropriate. Naturally, as anticipated modified Mercalli intensities increase, so too do the benefits associated with partial or total evacuation. The latter course of action would, for example, be appropriate if modified Mercalli intensities of IX, in highest risk microzones were expected.

Earthquake warnings cannot yet be issued with certainty. As a result, although the social stress likely to be caused by the anticipated seismic event can be calculated, as can that due to the prediction, actual stress prevented depends upon the accuracy of the forecasts. To successfully apply the methodology being developed here, the likelihood that the forecast will be correct must also be known. For this reason, it is suggested that earthquake warnings, like most predictions of precipitation, should be issued in probability form. A typical forecast, for example, might state that "there is a 50% probability that an earthquake capable of reaching modified Mercalli intensity VII on normal ground will occur during the next six months." If such predictions were issued, the stress benefit–cost ratio for different responses could easily be modified to reflect the degree of certainty of the forecast. In the case of such a prediction of the 100-year earthquake for the city of Victoria, anticipated social benefits could be halved, while the costs of response would remain unchanged. Appropriate official action would vary accordingly.

If warning systems are to operate effectively, the final responsibility for decisions must be very clearly defined. Ideally, a manual should be prepared outlining every step of the warning process. The procedures involved should then be approved by the governor, mayor, safety plan coordinator, or whomever bears ultimate political responsibility. Such a formalization reduces the personal stress on those individuals whose prime responsibility it is to run the network, making them less likely to err on the side of caution. However, manuals must leave room for initiative in unforeseen circumstances to avoid rigidity and inflexibility. In the final analysis it is far better to warn in error than to fail to warn when disaster does strike.

Transmission of Warning Messages

No warning system will function effectively if its messages, however logically arrived at, are ignored, disbelieved, or lead to inappropriate action. To avoid such pitfalls, warning messages must be designed and transmitted with the greatest of care.

All warnings must be specific, clearly identifying who is expected to respond. Unfortunately this has often not been the case, a problem that was dramatically illustrated by the warnings issued prior to Hurricane Carla's devastating impact in Lower Cameron Parish, Louisana in 1957 (Williams, 1964). Residents in this area received radio broadcasts from Lake Charles in the same state, which reassured them that there was little need for alarm. They therefore took no steps to protect themselves, while in fact they were in very grave danger. Unfortunately the reassuring messages had only been intended for Lake Charles, some 60 miles distant.

Warning messages should also include details of the expected time of impact of the disaster agent, in addition to the specific area threatened. When immediate action is required graphic descriptions of the damage that has already occurred or is likely to take place might be included in the message. McLuckie (1970) includes the following example in his study of disaster warning systems:

> A flood of major proportions is sweeping into the city. All valley residents should evacuate their homes immediately and go to high ground.

This message was reinforced with the following eyewitness account, issued by a Denver radio station:

> I've never seen anything like it, it's sweeping everything before it. A wall of water 20 feet high and about one-half mile wide is moving fast. There are trailers in it, houses, cows, trees. Bridges are snapping from the dikes. It's going to hit Denver and it's bad. My God!

While perhaps a little melodramatic, such a warning is likely to result in rapid evacuation. This is certainly better than the warning issued by the flood forecasters of the U.S. National Weather Service immediately prior to the large scale destruction in Wilkes-Barre. This merely stated that a 40-ft crest would reach the city in a few hours. Such a river height would inevitably cause enormous damage and since the "coolness" of the flood forecaster implied no such destruction, the warning was thought to be based on a miscalculation. Local officials therefore continued sandbagging the river, an utterly hopeless task in view of the magnitude of the approaching floodwaters (Cullen, 1978).

Messages must also be specific in terms of their consequences. For example, does a flood warning mean that residents should move their belongings onto the second floor or evacuate the area completely? Will a predicted epidemic be likely to cause mild illness or a painful death? Only if the projected consequences of the threat are carefully given can respondents make rational decisions. For this reason, the probability of occurrence should be realistically given, particularly when it is much higher than has been the case in earlier potential disaster situations. In 1964, for example, a tsunami warning was issued in Crescent City, California, 52 minutes before the arrival of the first small wave and two and a half hours before the fourth and more destructive wave hit the area. Despite the time available, few people evacuated the threatened area and little was done to reduce the damage potential. This was because past experience had taught local people that the "probable" seismic wave had, in fact, very little chance of occurring (McLuckie, 1970). As a result, at least 11 residents were killed and property was damaged more heavily than was necessary over a 29-block area.

It is also very important that prototype warning messages should be as completely prepared as possible. They can often be written except for gaps where specific details of time, probability of arrival, and scale of impact can be quickly added. It is also useful to prerecord a wide variety of television and radio tapes. These should be carefully labeled before distribution so that in the event of an emergency, instructions can be given to immediately play a specific warning. Press releases can also be prepared during the long periods of normalcy that precede disaster situations.

Since it is critical that such warning messages be believed, great thought should be given as to who should prerecord them. Typically, viewers and listeners are bombarded by hundreds of selling messages a week. They therefore tend to develop strong resistance to advertising. Thus it makes little sense to use personnel who are associated with such "hard selling" to record warning messages. Conversely, the safety plan coordinator may be unsuited because of an inability to communicate with the general public. Ideally, a series of warnings should be recorded by a variety of individuals, perhaps including the mayor, a priest, the chief of police, the safety plan coordinator, and local celebrities. These can then be shown in private to small groups of citizens and, on the basis of their reactions, the most suitable broadcast(s) selected.

It should be noted that many systems make distinctions between the issuing of hazard watches and hazard warnings. The former generally implies that conditions are those normally preceding disaster, that is, for example, an earthquake of large magnitude may have occurred beneath the ocean and may have generated a tsunami. A warning bulletin normally implies that there is positive proof or very great likelihood that the hazard is approaching the area. The public generally has considerable difficulty in differentiating between watch and warning messages. Examples of two types of warning and the process by which they were developed, issued by the Seismic Sea Wave Warning System (U.S. Department of Commerce, 1965) and Colorado's U.S. Forest Service avalanche warning program (Judson, 1976; Perla and Martinelli, 1976), follow to illustrate the principles involved.*

0344 Seismic sea-wave warning alarm sounds at the Honolulu Observatory. Requests issued for immediate readings from seismograph stations at College, Alaska; Sitka; Pasadena; Berkeley; Tucson; Tokyo; and Guam.

*References are to Greenwich Mean Time (GMT), or ZULU, as the "Z" in teletype messages indicates. Pacific Standard Time (which includes Alaska panhandle) is 8 hours behind GMT: Yukon Time (Yakutat north), 9; Alaska Time (most of central Alaska) and Hawaiian Standard Time, 10; and Bering Time, 11.

0452 Honolulu Observatory completes preliminary determination of location of earthquake epicenter: 61 N., 147.5 W.

0502 Honolulu Observatory issues first advisory via VAA and Defense Communications network: THIS IS A TIDAL WAVE ADVISORY. A SEVERE EARTHQUAKE HAS OCCURRED AT LAT. 61 N., LONG. 147.5 W., VICINITY OF SEWARD, ALASKA, AT 0336Z, 28 MAR. IT IS NOT KNOWN, REPEAT NOT KNOWN, AT THIS TIME THAT A SEA WAVE HAS BEEN GENERATED. YOU WILL BE KEPT INFORMED AS FURTHER INFORMATION IS AVAILABLE. IF A WAVE HAS BEEN GENERATED, ITS ETA FOR THE HAWAIIAN ISLANDS (HONOLULU) IS 0900Z, 28 MARCH ...

0530 Communications inoperative on Alaskan mainland. Honolulu Observatory issues second bulletin; this is an information bulletin, stating that it is not yet known whether a seismic sea wave has been generated, but providing all participants in the SSWWS with estimated times of arrival.

The observatory also requests inspection of tide records by observers at Unalaska, Kodiak, Adak, Sitka, Alaska; and Crescent City, California.

0555 Kodiak replies:
EXPERIENCE SEISMIC SEA WAVE AT 0435Z. WATER LEVEL 10-12 FT ABOVE MEAN SEA LEVEL. WILL ADVISE.

0630 Another message from Kodiak tide observer confirms existence of seismic sea wave.

Honolulu Observatory issues third bulletin: THIS IS A TIDAL WAVE/SEISMIC SEA WAVE WARNING. A SEVERE EARTHQUAKE HAS OCCURRED AT LAT. 61 N., LONG. 147.5 W., VICINITY OF SEWARD, ALASKA, AT 0336Z, 29 MAR. A SEA WAVE HAS BEEN GENERATED WHICH IS SPREADING OVER THE PACIFIC OCEAN. THE ETA OF THE FIRST WAVE AT OAHU IS 0900Z, 28 MARCH. THE INTENSITY CANNOT, REPEAT, CANNOT BE PREDICTED. HOWEVER, THIS WAVE COULD CAUSE GREAT DAMAGE IN THE HAWAIIAN ISLANDS AND ELSEWHERE IN THE PACIFIC AREA. THE DANGER MAY LAST FOR SEVERAL HOURS ... (estimated times of arrival are repeated).

0700 Tsunami reaches Tofino, B.C.

0708 Kodiak reports series of waves:
SEA WAVES AT 0435Z; 32 FT AT 0540Z; 35 FT AT 0630Z; 30 FT SEAS DIMINISHING, WATER RECEDING. EXPECT 6 MORE WAVES.

0739 First wave arrives at Crescent City. The wave is 3 feet high. Some evacuees return to danger area.

0750+ Four persons drown at Depoe Bay, Oregon.

0900 Tsunami reaches Hawaiian Islands. Damage slight at Hilo, with three restaurants and a house inundated; at Kahului, a shopping center is flooded.

0920 A 12-foot wave—probably the fourth—sweeps into Crescent City. This wave, and its successors, destroy or displace more than 300 buildings; 5 bulk gasoline storage tanks explode; 27 blocks are substantially destroyed; there are casualties.

1020 Tsunami reaches east coast of Hokkaido, Japan.

1038 Honolulu Observatory sends final bulletin. It is an all-clear for Hawaii; other participants in the SSWWS are advised to assume all-clear status two hours after their tsunami ETA unless local conditions warrant continuation of alert.

1035 Tsunami reaches Kwajalein.

1910 Tsunami reaches La Punta, Peru.

Avalanche warnings provide a further example of the correct procedure to be followed:

ZCZC
AVUS RWRC 100300
....AVALANCHE WARNING....BULLETIN NUMBER 3
IMMEDIATE BROADCAST REQUESTED
U.S. FOREST SERVICE FORT COLLINS COLORADO
ISSUED 8PM MST MONDAY FEBRUARY 9 1976

CENTRAL AND SOUTHERN COLORADO MOUNTAINS

THE AVALANCHE WARNING FOR THE COLORADO MOUNTAINS SOUTH OF A LINE FROM GRAND JUNCTION TO LEADVILLE REMAINS IN EFFECT AND IS VALID THROUGH WEDNESDAY FEBRUARY 11 1976.

DANGER FROM SNOWSLIDES IN PORTIONS OF THE SAN JUAN MOUNTAINS IS NOW EXTREME DUE TO VERY HEAVY SNOWFALL AND HIGH WIND. NUMEROUS LARGE AVALANCHES OCCURRED TODAY AND TONIGHT....MORE ARE EXPECTED. SIXTY AVALANCHES WERE REPORTED FROM THE WARNING AREA IN THE PAST SEVEN HOURS.

ALL MOUNTAIN TRAVELERS SHOULD USE EXTREME CAUTION IN THE WARNING AREA.

THE NEXT AVALANCHE WARNING BULLETIN WILL BE ISSUED TUESDAY AT 11AM OR EARLIER IF CONDITIONS WARRANT.

BACHMAN/JUDSON....USFS. FORT COLLINS COLORADO

ZCZC
AVUS RWRC 122300
AVALANCHE WARNING TERMINATION....BULLETIN NUMBER 9
IMMEDIATE BROADCAST REQUESTED
U.S. FOREST SERVICE FORT COLLINS COLORADO
ISSUED 4PM MST THURSDAY FEBRUARY 13 1976

SOUTHERN AND CENTRAL COLORADO MOUNTAINS

THE AVALANCHE WARNING FOR THE SOUTHERN AND CENTRAL COLORADO MOUNTAINS....SOUTH OF A LINE FROM GRAND JUNCTION TO LEADVILLE....HAS BEEN TERMINATED. AVALANCHE DANGER HAS MODERATED IN THE AREA BUT BACKCOUNTRY USERS SHOULD CONTINUE TO USE CAUTION IF TRAVELING IN STEEP TERRAIN.

APPROXIMATELY 175 AVALANCHES WERE REPORTED IN THE WARNING AREA SINCE LAST MONDAY MORNING. A FLIGHT OVER THE AREA TODAY SHOWED THAT SEVERAL HUNDRED MORE AVALANCHES RAN IN REMOTE AREAS BUT, OF COURSE, WENT UNREPORTED.

WILLIAMS/BACHMAN....USFS. FORT COLLINS COLORADO

A wide variety of channels and media should be used to distribute warning messages. Individuals are more likely to respond if they receive information from several sources, both formal and informal. Apart from the normal means of communication such as telephone fan-outs, radio, and television broadcasts, thought might be given to utilizing supermarket music systems, police, fire, taxi, truck, and citizen band radios and flags, sirens, whistles, flares, and door-to-door volunteers. Naturally where nonspecific warnings such as sirens are used, the public must have been educated about their significance. Thought might also be given to equipping the residences of individuals living in high risk areas or in positions of authority with disaster units which can be activated via satellite to play the appropriate prerecorded warning message. Electronic display boards, such as those used to warn motorists of avalanche hazards on British Columbian highways, can be induced to provide the necessary warning by dialing a telephone code. These aids have considerable potential for use in areas subjected to flash floods, tsunamis, forest fires, and many other hazards, particularly when the speed of onset is rapid.

The setting of warning messages is also very critical. Clifford (1955), for example, records the unfortunate case of Piedras Negras, a Mexican town threatened by the flooding Rio Grande. Since disaster was imminent, two loudspeaker cars were coopted from the local cinema to supplement four official units. It is said that one of these vehicles cruised through the city repeating the following message:

> An all-time record flood is going to inundate the city. You must evacuate immediately. (Pause) The ________________ theater is presenting two exciting features tonight. Be sure to see these pictures at the ________________ theater tonight.

Such a warning message is unlikely to result in anything but total bewilderment. A similar problem often occurs when warning messages are inserted into normal television and radio programming. If, for example, the Super Bowl were to be interrupted by a brief announcement of impending disaster and then the game broadcast resumed, it would be very easy for viewers to disregard the warning on the grounds that had it really been significant, normal broadcasting would have ceased. For this reason, arrangements must be made for emergency program packets to supersede normal broadcasting. Otherwise, much of the impact of the warning will be dissipated by the normality of its setting.

Interpretation of Warning Messages and Action by Recipients

In the early hours of 23 May 1960 a seismic sea wave, generated by earth movement off the coast of Chile, inundated the lower parts of Hilo, Hawaii. Despite the fact that authorities in the area had at least 10 hours advance notice, 61 people were killed, several hundred injured, and 500 homes and their contents destroyed (Lachman, Tatsuoka, and Bank, 1961). Soon after this disaster, 327 survivors were interviewed, in an attempt to discover why they had not evacuated the threatened area, in spite of prolonged warning. It was found that, although the seismic sea wave siren, which had sounded for a 20-minute period more than 4 hours prior to the impact of the first wave, had been heard by 95% of the survivors, only 40% actually evacuated their homes. Further questioning revealed that the threatened population had been very confused about the significance of those sirens. Numerous erroneous meanings had been attached to this signal to evacuate; these included a belief that the sirens were an alert, a preliminary signal preceding official evacuation, a notice to await further instructions or to make preparations to leave the area.

Such confusion could only lead to a variety of erroneous responses. Forty-four of those interviewed disregarded the sirens completely either because of lassitude, age, infirmity, or since they considered their own home or workplace to be completely safe. A further 131 survivors had been more concerned by the sirens but had awaited further instructions from the radio or television, or aid in evacuation or for a more official verbal notification of the need to leave the area. Such a wait-and-see response has also been reported from many other disaster threatened areas (Hughes, 1973; Haas *et al.*, 1976).

Given the confusion caused by the operation of the tsunami warning system in Hilo, it was perhaps not surprising that when the first seismic wave actually struck, 43% of those who did not evacuate admitted to being asleep, while a further 48% were at home but still awake. Most surprising of all, 14 survivors had gone down to the shore, specifically to watch the tsunami arrive.

The way in which individuals and families respond to disaster warnings has been studied by both sociologists and psychologists (Killian, 1952; Fritz, 1957; Drabek, 1969). Three important generalizations can be drawn from this research. First, even though a wide variety of people may be listening to the same warning message, everybody hears and believes different things. Second, people respond to warnings on the basis of how what they hear encourages them to behave. Third, individuals are stimulated differently depending on who they are, who they are with, and who and what they see (Mileti, 1975). Table 6.2 attempts to illustrate some of these factors in more detail. Response by organizations to disaster warnings is greatly influenced by their bureaucratic structure. Strongly established patterns of centralized decision making may delay action prior to disaster agent impact. This seems particularly true if the required reaction to warning drastically disrupts the normal activities of the organization (Mileti, 1975). Similarly, rural populations have been found to be more reluctant to evacuate than urban populations.

It is clear from this complexity that the behavior of those at risk cannot be easily predicted. It reflects present activity, personal prejudices, beliefs, education, experience, and health. Response can be improved, however, by the type of educational programs already discussed. The relative unpredictability of public reaction also necessitates the next step in the design of warning systems, the planning for feedback.

Feedback to Issuers of Warnings

Information should flow in both directions through warning systems, passing both to and from those threatened. Such feedback permits operators of warning networks to assess the impact of their messages. If those at risk are not responding as anticipated, new information can be provided and further actions taken to improve the situation. Although some hazards, such as tornadoes, have such a rapid speed of onset that response monitoring is largely precluded, most disaster agents allow it.

From the preceding description of the events in Hawaii on 23 May 1960 it is apparent that the seismic sea wave caused a large number of unnecessary deaths, injuries, and needless property damage. These losses occurred because of a malfunctioning of the area's tsunami warning system which did not include essential feedback from the users of its service. In the 10 hours between the generation of the tsunami off the coast of Chile and its arrival in Hawaii, the threatened sections of Hilo could have been completely evacuated, valuable possessions moved, and roadblocks set up to prevent the premature return of residents or the arrival of thrill seekers. That this did not happen was largely due to a failure, on the part of those issuing the tsunami warnings, to recognize the confusion and lack of rational response that their actions had created.

Except in the case of those hazards where very rapid speed of onset prevents feedback, every warning system should be designed to facilitate a two-way flow of information. Ideally, members of police forces, the military, other emergency services, and concerned members of the public should be trained as contacts, long before any disaster threatens. Certain members of the official warning agency should be given the specific task of coordinating their efforts and ensuring that their information is received and acted upon. Special communication channels, such as citizen band or taxi cab radios, should be established to make certain that user response to warnings can be closely and rapidly monitored. It is also very important that those members of the public, trained to play this liaison role in the event of imminent disaster, should be respected community leaders. Such status makes it more likely that their advice will be followed. These individuals must also receive prior training and be kept well informed of the evolving disaster situation, so allowing them to provide accurate information to a concerned public.

Table 6.2. Factors Influencing Response to Disaster Warnings

1. Any warning message broadcast, especially the early ones, will be accepted at face value only by a minority of the recipients. Most will engage in confirmation efforts for a time.
2. The more warning messages received by an individual, the fewer the attempts at warning confirmation.
3. The closer a person is to the target area of a warning, the higher the incidence of face-to-face communication and the larger the number of sources used in confirmation attempts.
4. Warnings from official sources (police, state patrol, fire department) are more likely believed.
5. Message content per se influences belief. The more accurate and consistent the content across several messages, the greater the belief.
6. The more personal the manner in which a message is delivered, the more it will be believed.
7. Belief in eventual impact increases as the number of warnings received increases.
8. The recipient's sense of the sender's certainty about the message is important to belief.
9. Message believability is related to what happens in the confirmation process. The response of official sources to questions which call for validation, corroboration, or refutation helps determine believability.
10. A person is more likely to believe a warning of impending danger to the extent that perceived changes in his physical environment support the threat message.
11. Persons who see others behaving as if they believe a warning to be valid are themselves more likely to believe the warning.
12. Past experience may render current warnings less credible if disaster is not part of experience, or more credible if disaster is part of experience.

Unfortunately, in any free society there are always a few individuals who appear bent on self-destruction. These included the 14 lucky survivors who went down to the beach at Hilo to watch the Hawaiian tsunami arrive. Similarly, during Hurricane Camille, some 30 fatalities had refused to evacuate, preferring to attend a "hurricane party" in a building which was later demolished by the storm. Local authorities might consider compulsory evacuation under such circumstances. Possible drastic measures that could be used include arrest for disturbing the peace, or making children wards of the court because of parental neglect. Naturally, local governments must examine their powers of compulsion and associated legal responsibilities well in advance of any imminent disaster. In addition to those who simply refuse to leave threatened areas, there are many who cannot do so without aid. These include the very young, elderly, and infirm.

Table 6.2. (*continued*)

13. The closer a person is to the target area of warning, the more rumors he will hear and the less accurate will be his understanding of the character of the forecast events.
14. Persons do not readily evacuate on the basis of the first warning received, and the number of warnings received thereafter is directly related to evacuation.
15. As the warning message increases in its accuracy, and/or information about survival choices, and/or consistency with other warnings, and/or clarity about the nature of the threat, the probability of adaptive response increases.
16. Whether or not a person takes action depends on his belief in the warning message. But even if he believes, he may fail to take *adaptive* action due to his misinterpretation of the meaning of the message content.
17. Evacuation tends to be a family phenomenon. The best way to accomplish evacuation appears to be repeated authoritative messages over broadcast media which stimulate discussion within the family and lead to evacuation (if it is going to happen at all).
18. Persons receiving face-to-face warnings in a family setting from authorities are more likely to evacuate.
19. Persons with recent disaster experience are more likely to take protective actions.
20. Perceived amount of time to disaster impact is also important.
21. Belief that impact could occur at the location from which a person may be about to evacuate is critical.
22. Older persons are less likely than the young to receive warnings regardless of warning source, and less likely to take protective actions.
23. Regardless of the content of a warning message, people tend to define some potential impact in terms of prior experience with *that specific* disaster agent.

Source: Haas (1973) and Mileti (1975).

Transmission of Further Warning Messages

Warning messages should be repeated as frequently as possible since waverers may respond to the fourth or fifth message, if not to earlier warnings. This is more likely to occur if the decreasing time available to respond is stressed in later messages. The need for such confirmation was demonstrated during the Darwin cyclone disaster (Haas, Cochrane, and Eddy, 1976). It is in this context that feedback is most valuable since it permits changes in the nature and urgency of the information relayed. Later warnings can be structured so as to reduce existing misconceptions and prevent any further undesirable responses.

The transmission of additional postdisaster warnings may be hindered by damage to normal communications systems. In the Wyoming Valley flood the telephone exchange was badly affected and its use for subsequent warnings largely precluded (Cullen, 1978). Similarly, potential or actual disasters commonly lead to convergence, an enormous movement of information, personnel and materials into the stricken area. This process often interferes with the transmission of subsequent warnings. Telephone networks may be overloaded by calls to reassure friends and relatives of safety or to seek information about injured or missing persons. To ensure that this does not lead to a breakdown in the warning system, plans must be made long before disaster strikes to reserve certain communication channels for official use.

Unfortunately, many disasters involve some degree of secondary threat which may prove more damaging than the initial event. Seismic activity, for example, often causes large scale fires. Floods and earthquakes may also damage sewage disposal and drinking water supplies, making the spread of dysentery, typhoid, and other water-borne diseases very possible. Numerous other secondary threats from escaping chemicals, gases, or wastes are also possible. Any new danger must become the subject of supplementary warnings.

Issuing of All-Clear Messages

There are many myths about the nature of threat (Vitaliano, 1973). During the 1964 Alaskan tsunami, three Aleuts were killed in Kaguyak, because of their belief that the second wave was always the largest (Engle, 1966). After this had passed without damage to their property they returned, only to be killed by the larger third seismic wave.

It is only to be expected that homeowners would prefer to spend time in their own residences rather than in evacuation centers or with friends. For this reason it is essential that subsequent warning messages stress that it is not yet time to return and that, when the danger has passed, an all-clear will be issued. Because of the wide-spread occurrence of premature returns to evacuated areas, it is essential that perimeters are guarded to prevent unauthorized reentry.

Hindsight Review

Disaster warning systems can always be improved. The necessity for such changes becomes most apparent during actual disaster situations. Therefore, it is essential that the operation of the warning system itself be carefully recorded. This may appear an unnecessary chore to those involved in dealing with the demands of escalating threat, but it is very necessary if the operation of the system is to be improved in future disaster situations.

This process of hindsight review is likely to be most successful if detailed records are kept of the wording and time of arrival of all data, watch or warning messages, the system's response to them, and public reaction to this information. It is therefore useful if log books, tape recorders, and videotape equipment have been readied by the safety committee for this purpose well in advance of any danger. Such material is also very beneficial in the subsequent production of public education programs and in the training of new personnel.

Once the risk of disaster has passed, or damage and injury satisfactorily dealt with, all those involved in the operation of the warning system should be debriefed. Discussions should be held between those involved, to record in detail problems experienced and reasons for them. To avoid interdepartmental and personal conflict, the chairman of such hindsight review sessions should stress the value of what was achieved, and attempt to ensure that discussions focus upon possible improvements to the system rather than criticism. Such a debriefing was successfully undertaken after the 1971 Regina flood. This particular exercise led to the identification of several problems, such as the lack of personnel control in some areas and difficulty with sanitation, first aid, food, and traffic control. It was also found that because of the use of walkie-talkies by children, some important official communications had been jammed. The Regina emergency operations center and telephone fan-out system was found to have operated quite well (Walker, 1971). Conflicts may arise when agencies see this review process as a mechanism for arguing for larger budgets, greater power, and additional personnel as was the case after the Baldwin Hills dam burst (Anderson, 1964).

In extreme cases where the agencies involved are unable to conduct the review process unaided, outside consultants should be hired to undertake an independent assessment. This may be particularly necessary when the warning system has malfunctioned.

It is also very useful to undertake an interview and questionnaire survey among those served by the network. This should seek to determine whether the warnings were heard, and when and what actions were taken as a result. More detailed information might be solicited from those who failed to respond as expected, so that the content of future messages can be more effective.

Testing the Revised System

The chief purpose of a hindsight review is to achieve improvement through adverse experience. This process should lead to beneficial changes in the operation of the warning network. It cannot be assumed that all such proposed alterations will in fact prove to be successful, and it is essential that, after changes have been made, the system is tested once more. Such disaster simulations, undertaken as step five in establishing the network, must be evaluated and acted upon where necessary.

References

Anderson, W. 1964. *The Baldwin Hills, California Dam Disaster.* Research Note No. 5, The Disaster Research Center, Ohio State University, Columbus, Ohio.

Burgess, D. W., L. D. Hennington, R. J. Doviak, and P. S. Ray. 1976. Multimoment Doppler display for severe storm identification. *Journal of Applied Meteorology,* **15(12)**:1302–1306.

Burton, I., R. W. Kates, and G. F. White. 1978. *The Environment as Hazard.* Oxford University Press, New York.

Clifford, R. A. 1955. *Informal Group Actions in the Rio Grande Disaster.* National Academy of Sciences-National Research Council, Washington, D.C.

Coe, R. S. 1971. Earthquake prediction program in the People's Republic of China. *EOS Transactions of the American Geophysical Union,* **52(12)**: 941–943.

Cullen, J. M. 1978. *A Comparison of Lifeline System Vulnerability in Two Large Regional Disasters: The Wyoming Valley Flood and the Projected Puget Sound Earthquake.* Report prepared for the Federal Disaster Assistance Administration, Region Ten, Seattle, Washington.

Dacy, D. C., and H. Kunreuther. 1969. *The Economics of Natural Disasters: Implications for Federal Policy.* Free Press, New York.

Drabek, T. E. 1969. Social process in disaster: Family evacuation. *Social Problems,* **16**:336–349.

Dunn, G. E., and B. I. Miller. 1964. *Atlantic Hurricanes.* Louisiana State University Press, Baton Rouge, Louisiana.

Engle, E. 1966. *Earthquake: The Story of Alaska's Good Friday Disaster.* John Day, New York.

Erb, J. H., and P. J. Wall. 1972. Tsunami warning in British Columbia. *EMO National Digest,* **12(1)**:1–3, 22.

Flood, M. 1976. Nuclear sabotage. *Bulletin of the Atomic Scientists,* October 1976, pp. 29–36.

Foster, H. D. 1976. Assessing disaster magnitude: A social science approach. *The Professional Geographer,* **XXVIII (3)**:241–247.

Foster, H. D., and R. Carey. 1976. The simulation of earthquake damage. *In:* H. D. Foster (Ed.), *Victoria Physical Environment and Development, 12,* Western Geographical Series. University of Victoria, Victoria, B.C., pp. 221–240.

Foster, H. D., and V. Wuorinen. 1976. British Columbia's tsunami warning system: An evaluation. *Syesis,* **9**:115–122.

Frank, N. L., and S. A. Hussain. 1971. The deadliest cyclone in history. *Bulletin American Meteorological Society,* **52(6)**:438–444.

Friedsam, H. J. 1960. Older persons as disaster casualties. *Journal of Health and Human Behaviour,* **1**:12–21.

Fritz, C. E. 1957. Disasters compared in six American communities. *Human Organization,* **16(Summer)**:6–9.

Gruntfest, E. C. 1977. *What People Did During the Big Thompson Flood.* Working Paper No. 32, Institute of Behavioral Science, University of Colorado, Boulder, Colorado.

Haas, J. E. 1973. What Every Good News Director Ought to Know About Disaster Warnings. Paper presented at the Radio and Television News Directors' Association Session on Natural Disasters, Seattle, Washington, October 10–11. Institute of Behavioral Science, University of Colorado, Boulder, Colorado.

Haas, J. E., H. C. Cochrane, and D. G. Eddy. 1976. *The Consequences of Large-Scale Evacuation Following Disaster: The Darwin, Australia Cyclone Disaster of December 25, 1974.* Working Paper No. 27, Institute of Behavioral Science, University of Colorado, Boulder, Colorado.

Hodgson, J. H. 1964. *Earthquakes and Earth Structure.* Prentice-Hall, Englewood Cliffs, New Jersey.

Hughes, E. 1973. Pink was the color of death. *Reader's Digest,* December 1973, pp. 184–190.

Judson, A. 1976. Colorado's avalanche warning program. *Weatherwise,* **29(6)**:268–277.

Kiersch, G. A. 1965. The Vaiont Reservoir disaster. *Mineral Information Service,* **18(7)**:129–138.

Killian, L. M. 1952. The significance of multi-group membership in disaster. *American Journal of Sociology,* **57**:309–314.

Lachman, R., M. Tatsuoka, and J. W. Bank. 1961. Human behavior during the tsunami of May 1960. *Science,* **133**:1405–1409.

Lewis and Clark Law School. 1977. *Legal Constraints on the Planning and Development of Disaster Home Warning Systems.* Report to the National Science Foundation Research Applied to National Needs Directorate.

McDavid, J., and H. Harari. 1968. *Social Psychology.* Harper and Row, New York.

McLuckie, B. F. 1970. *The Warning System in Disaster Situations: A Selective Analysis.* Ohio State University, Disaster Research Center Report Series No. 9, prepared for Office of Civil Defense, Office of Secretary of the Army, Washington, D.C.

Mileti, D. S. 1975. *Natural Hazard Warning Systems in the United States: A Research Assessment.* Program on Technology, Environment and Man, Institute of Behavioral Science, University of Colorado, Boulder, Colorado, 99pp.

Moore, H. E. 1958. *Tornadoes over Texas.* University of Texas Press, Austin, Texas.

Moore, H. E., and F. R. Crawford. 1955. *Waco–San Angelo Disaster Study: Report of Second Year's Work.* Unpublished report, University of Texas, Austin, Texas.

Morton, D. 1978. *Directory of Sources for Films and Other Visual Materials on Natural Hazards and Their Mitigation.* Institute of Behavioral Science, University of Colorado, Boulder, Colorado, 9pp.

Mussari, A. J. 1974. *Appointment with Disaster.* Northeast Publishers, Wilkes-Barre, Pennsylvania.

Owen, H. J. 1977. *Guide for Flood and Flash Flood Preparedness Planning.* U.S. Department of Commerce, National Oceanic and Atmospheric Administration, National Weather Service, Disaster Preparedness Staff Report.

Perla, R. I., and M. Martinelli, Jr. 1976. *Avalanche Handbook.* U.S. Department of Agriculture, Forest Service, Agriculture Handbook 489.

Riley, J. A. 1971. *Disaster-Storm Ahead.* The Hogg Foundation for Mental Health, Austin, Texas.

Roberts, W. O. 1972. We're doing something about the weather! *National Geographic* **141(4)**:518–555.

Schwartz, G. 1979. Tornado! *Journal of Civil Defense,* **XII(2)**:10–13.

Sewell, W. R. D., and H. D. Foster. 1975. Disaster Scenario: British Columbia's Tsunami Warning System. Paper presented at the Environmental Hazards and International Development Session, Pacific Science Congress, Vancouver, British Columbia, August 1975.

Shaw, M. 1978. Smoke detectors can save your life. *Reader's Digest,* **113**:675–679.

Spaeth, M. G., and S. C. Berkman. 1967. *The Tsunami of March 28, 1964 as Recorded at Tide Stations.* U.S. Department of Commerce, Environmental Science Services Administration, Coast and Geodetic Survey, Technical Bulletin No. 33.

Stirton, A. M. 1971. Emergency public information planning in Canada. *EMO National Digest,* **11(5)**:1–6.

Susquehanna River Basin Commission. 1976a. *Planning Guide Self-Help Flood Forecast and Warning System Swatara Creek Watershed, Penna.*

Susquehanna River Basin Commission. 1976b. *Neighborhood Flash Flood Warning Program Manual.*

Torres, K., and P. Waterstone. 1977. *Information Services for Natural Hazards Research—Organizations, Periodicals, Newsletters, and Reference Sources.* Natural Hazards Research and Applications Information Center, Institute of Behavioral Science, University of Colorado, Boulder, Colorado.

Tsuruta, E. C. 1963. *Tsunami—Its Nature and Counter-Measures Against It.* Report prepared for the Japanese Port and Harbor Technical Research Institute, Ministry of Transport.

U.S. Department of Commerce. 1965. *Tsunami: The Story of the Seismic Sea-Wave Warning System.* U.S. Department of Commerce, Coast and Geodetic Survey.

U.S. Department of Commerce. 1971. *Communication Plan for Tsunami Warning System.* National Oceanic and Atmospheric Administration, National Ocean Survey.

U.S. Office of Emergency Preparedness. 1972. *Disaster Preparedness.* Report to Congress, Executive Office of the President, 184 pp.

Vitaliano, D. B. 1973. *Legends of the Earth: Their Geologic Origin.* Indiana University Press, Bloomington, Indiana.

Walker, H. 1971. Regina: EMO response to flood. *National Digest,* **11(3)**:27–32.

White, T. A. 1969. Satellites gave warning of midwest floods. *National Geographic,* **136(4)**:574–592.

White, W. R. H. 1966. The Alaska earthquake—Its effects in Canada. *Canadian Geographical Journal,* **LXXII**:210–219.

Williams, H.B., 1964. Human factors in warning-and-response systems, in G. H. Grosser, *et al.*(eds.) *The Threat of Impending Disaster: Contributions to the Psychology of Stress.* MIT Press, Cambridge, Massachusetts, 80-96.

Working Group on Earthquake Hazards Reduction. 1978. *Earthquake Hazards Reduction: Issues for an Implementation Plan.* Office of Science and Technology Policy, Executive Office of the President, Washington, D.C., 231 pp.

7
Disaster Plans

The Moving Finger writes; and, having writ,
Moves on: nor all thy Piety nor Wit
Shall lure it back to cancel half a Line,
Nor all thy Tears wash out a Word of it.

The Rubá' iyát of Omar Khayyám
Edward Fitzgerald (1809–1883)

The Case for Preparation

Local authorities differ in the efficiency with which they manage such routine responsibilities as water supply and refuse disposal. This degree of competence is reflected in both the quality and cost of the services provided. Inefficiency in disaster planning, however, has far more serious repercussions, often resulting in numerous avoidable deaths and injuries. Each municipality should be prepared to respond with authority to threat and be able to cope unaided with the initial impact of any disaster. This task is not beyond the capabilities of most local authorities, since they generally have the facilities on hand to deal temporarily with the demands generated by disaster (Healy, 1969).

The scale of these requirements can be estimated from examination of experience elsewhere. The highest death toll resulting from a single North American disaster in the twentieth century was the 6000 fatalities that occurred during the 1900 Galveston, Texas, hurricane storm surge. Only four other North American disasters have resulted in over one thousand fatalities in this time period. Much more typically, even major agents of destruction cause less than 200 deaths and perhaps 1000 associated casualties. Realistically, then, it is this more typical emergency that a North American disaster plan should seek to mitigate. Naturally, much larger losses can almost always be postulated in extreme social, geophysical, or meteorological conditions but contemplating them often leads to the conclusion that they are too catastrophic to allow rational planning. In consequence, such exercises may lead to a dangerous inertia on the part of local authorities.

Emergencies are characterized by four distinguishing features, an urgent need for rapid decisions accompanied by acute shortages of the necessary trained personnel, materials, and time to carry them out effectively. Even under normal conditions decision making is a complicated process which involves such steps as problem definition, the ranking of objectives, the establishing of criteria, and the recognition of constraints. Alternative strategies for achieving objectives must be evaluated, programs compared, and the consequences of particular actions estimated. Once these stages in the decision-making process have been carried out, plans must be drawn up, implemented, and their related performances and impacts evaluated (Mitchell, 1971). The complexity of this process is increasing and under normal circumstances several years may elapse between problem identification and plan implementation. In emergency situations such a process is clearly impractical. Victims trapped beneath fallen buildings, on the roofs of burning highrises, or caught in rapidly submerging residences cannot afford the luxury of bureaucratic procrastination. Similarly, social services such as hospitals or welfare agencies may suddenly be called upon to provide aid on a scale far beyond normal. Without careful preplanning, breakdowns in emergency decision making inevitably leads to inferior performance, heightening disaster losses.

Since under normal circumstances municipal decision making is a complicated and protracted process, a well-designed disaster plan should seek to identify the types of decisions that are likely to arise during an emergency situation. Procedures for dealing with them must be established beforehand and the process of government streamlined to allow immediate yet optimum responses when disaster strikes. The safety plan coordinator and his committee should play a major role in making sure that such a document is prepared, circulated, and tested.

The Content of a Disaster Plan

While every community has special needs and capabilities which should be reflected in its planning, it is possible to provide a general disaster plan format. This is outlined in Table 7.1 and discussed in more detail in the following section.

Policy Statement

No disaster plan is likely to prove effective if it does not have the strong support of senior members of the government for which it is being prepared. This cannot simply take the form of a casual, tacit agreement that disaster planning is a useful exercise, but rather requires a forceful demonstration of commitment. Personal involvement by the mayor and aldermen is required to demonstrate to employees and the general public that concrete support in terms of finances, equipment, and personnel

Table 7.1. Typical Contents of a Disaster Plan

1. Policy statement on value of disaster planning by chief executive officer
2. Legislative authority for the design of the disaster plan and for the steps it contains
3. Aims of the plan and conditions under which it comes into force
4. Assessment of community disaster probabilities
5. Disaster scenarios
6. Relationships with other levels of government, particularly emergency-related agencies
7. Authority organization chart
8. List of names, addresses, and telephone numbers of all relevant agencies, their heads and deputies
9. Operation of warning systems:
   (1) types of warnings
   (2) distribution
   (3) obligations on receiving warnings.
10. Preimpact preparations:
    (1) relationships between type of disaster agent and necessary preparations
    (2) responsibilities of different agencies
    (3) location of greatest risk sites
11. Emergency evacuation procedures:
    (1) conditions under which evacuation is authorized
    (2) routes to be followed and destinations
    (3) accommodating the special needs of the elderly, ill, or institutionalized
12. Shelters:
    (1) locations
    (2) facilities
13. Disaster control center and subcenters:
    (1) location(s)
    (2) equipment
    (3) operation
    (4) staffing
14. Communications
15. Public information
16. Search and rescue:
    (1) responsibilities
    (2) equipment
    (3) areas most likely to require servicing
17. Community order
18. Medical facilities and morgues:
    (1) location
    (2) transportation
    (3) capacity
    (4) facilities
19 Restoration of community services:
    (1) order of priorities
    (2) responsibilities

20. Protection against continuing threat:
    (1) the search for secondary threats
    (2) actions to be taken if discovered
21. Continuing assessment of total situation:
    (1) responsibilities
    (2) distribution
22. Reciprocal agreements and links with other municipalities
23. Testing the plan:
    (1) disaster simulations
    (2) simulation evaluations
24. Revision and updating of the plan
25. Plan distribution

---

will be immediately forthcoming to design and, if necessary, implement the disaster plan. This commitment should include the passage of special enabling legislation which obligates the region and supportive public statements by the major or other senior officials that leave no room for misinterpretation. An example of such support by the mayor of Santa Rosa (1972) is presented below:

OFFICE OF THE MAYOR
THE CITY COUNCIL OF SANTA ROSA

February 29, 1972

Foreword to EMERGENCY SERVICES PLAN

The protection of life and property and the alleviation of suffering and hardship caused by disasters are fundamental responsibilities of civil government. Your City Officials are cognizant of these responsibilities and are determined that they shall be met.

Floods, fires, earthquakes, plane crashes and other major disasters occur at unexpected times and places throughout the world. Some of these happen every year in California and we may expect them to occur in the future. We are aware also that enemy weapons capable of mass destruction could strike our city with little or no warning.

For us to effectively cope with disaster conditions of any nature, requires careful planning and organizing, proper management of our resources and, a continuing training effort. This Emergency Services Plan, prepared by our Emergency Service Staff, is designed to provide for the use of public and private resources as necessary to cope with a wide variety of emergency situations.

Emergency Service is based directly on civil government, organized and augmented by the private sector as necessary to combat the severe and abnormal conditions which arise during emergencies. All Santa Rosa City Officials and Employees are members of the Emergency Service Organization.

The City Manager is the Director of the City of Santa Rosa

Emergency Service Organization and is charged with supervising the city's emergency readiness program. Although this plan is one step toward preparedness, continued awareness, training and personal readiness, particularly on the part of government employees, is necessary if we are to respond effectively under the varying conditions of disaster. Equally important is the understanding, support and cooperation of our citizens.

John Downey, Mayor
Santa Rosa City Council

Legislative Authority

In addition to such supportive statements, any plan should include details of its legislative base. For example, the *Emergency Operational Plan* (Winnipeg, 1964) of the Metropolitan Corporation of Greater Winnipeg contains the following introductory statements:

(a) Section 178(1) of Part X of the Metropolitan Winnipeg Act provides that:

> On the coming into force of this section, the corporation shall do everything that, in the opinion of the metropolitan council, is possible, practicable, and within the means of the corporation, to reduce the consequences to civilian persons and property in the metropolitan area of enemy action, flood or other disaster, *to cope with which is beyond the capacity of the regular services of area municipalities,* and to provide continuity of local government; and, without restricting the generality of the foregoing, but subject to anything done under section 8 of the Civil Defence Act, the corporation shall, in so far as the council deems it to be reasonable, practicable, and within the means of the corporation, plan, organize, take defensive or precautionary measures, including the strengthening of buildings and the construction and repair of dykes, and provide fire fighting, first aid, and evacuation services, and repair and restore public services and train personnel for any one or more of those things.

(b) By-Law 87 of the Metropolitan Corporation of Greater Winnipeg establishes an Emergency Measures Committee and an Emergency Measures Planning Committee to carry out the functions set out in the Metropolitan Winnipeg Act.

Similarly, the *Chatham-Savannah Civil Defense Operational Survival Plan* (1971) lists ten pieces of legislation which support its mission. These range from Public Law 920, 81st Congress to the Georgia Civil Defense Act of 1951 as amended and the Savannah Resolution, 24 May 1961.

Such detailed citations from legislation make it abundantly clear that the municipal disaster plan is a binding official document with the full sanction of local, regional, and federal governments.

Aims and Implementation

Every disaster plan should also include a succinct description of the purpose of the document and the conditions under which it becomes operative. Although these will vary from municipality to municipality, an example of such factors are provided below for the Santa Rosa emergency plan:

Purpose

This document with its associated annexes and standard operating procedures constitutes the City of Santa Rosa Emergency Services Plan. Its purposes are to:

(1) Provide the basis for the conduct and coordination of operations and the management of critical resources during emergencies.

(2) Establish a mutual understanding of the authority, responsibilities, functions and operations of civil government during emergencies.

(3) Provide a basis for incorporating into the City emergency organization non-governmental agencies and organizations having resources necessary to meet unforeseeable emergency requirements.

Activation of Emergency Services Plan

(1) This plan shall become operative:

(a) Automatically by the existence of a war-caused STATE OF WAR EMERGENCY as defined by the California Emergency Services Act.

(b) On order of the Mayor, the Director of Emergency Services, provided that the existence or threatened existence of a local emergency has been proclaimed in accordance with the provisions of the Emergency Services Ordinance of the City of Santa Rosa.

(2) The Director of Emergency Services is authorized to order the mobilization of the City emergency organization or any portion thereof as required to provide for increased readiness in event of the threatened existence of an emergency and prior to the activation of this plan.

Assessment of Community Disaster Probabilities

For a truly effective disaster plan to be prepared the safety plan coordinator and his committee must be cognizant of the hazards facing the community. The wide range of potential sources of relevant data has already been discussed, so too have the use of scale and mathematical modeling and the application of the Delphi technique in identifying hazards. Expert position papers can also be commissioned to pinpoint such threats. This latter alternative has been carried out in San Diego where at least seven hazards were identified (Abbott and Victoria, 1977). Once the most potentially destructive disaster agents have been recognized, they can then be used as a focus for the body of the community disaster plan.

Attention can then be turned to the agent-generated and response-generated demands associated with each hazard. The *Natural Emergency Plan* of the Ottawa-Carleton Emergency Measures Organization (1970), for example, includes an appendix in which each probable disaster agent is analyzed in isolation. Attention is drawn to its probable major effects, agency responsibilities, their possible actions at the scene of impact, and the required location and nature of necessary equipment. Two examples of this approach are provided in Table 7.2.

Table 7.2. Natural Emergency Plan: Types of Emergencies and Departmental Responsibilities

1. *Aircraft crash (urban) outside of airport*

A. Possible major effects

1. Casualties	7. International implications.
2. Deaths	8. Special cargo problems
3. Fire	9. Sudden hospital requirements
4. Explosion	10. Disruption of traffic
5. Damage to property	11. Disruption of utilities
6. Nuclear cargo problems	12. Convergence

B. Potential actions at the scene	Agency responsible
1. Establish an emergency headquarters	Municipal government/EMO
2. Establish adequate communications	EMO
3. Define a working area and establish a control perimeter	Operations control group
4. Secure disaster scene for subsequent investigation	Police
5. Eliminate hazards from damaged utilities	Engineering/utilities
6. Provide auxiliary power and lighting	Engineering/utilities/EMO
7. Rescue	Engineers/EMO
8. Fire fighting	Fire department
9. Notify hospitals of casualties including number and type	Medical/police/EMO
10. Establish a casualty clearance area and temporary morgue if required	Medical/coroner/ambulance service
11. Removal of casualties	Emergency ambulance service

12. Establish crowd control	Police
13. Establish traffic control	Police
14. Protect property and valuables	Police
15. Establish a news release system	Operations control group
16. Coordination and administration of incoming aid	Operations control group
17. Mobilize necessary manpower and equipment	Operations control group
18. Establish an inquiry service	Airlines/police/emergency welfare
19. Disposition of nuclear and special cargoes	Police/industry/radiation protection division

C. Equipment	Source
1. Fire fighting and rescue equipment	Fire/engineering/EMO
2. Ambulances	Emergency ambulance service
3. Communication equipment	Police/fire/EMO
4. Auxiliary lighting	Engineering/utilities/EMO
5. Barricades	Engineering
6. Mobile public address equipment	Police/fire/EMO/radio stations

2. *Dangerous Gases and Liquids*

A. Possible major effects

1. Casualties
2. Deaths
3. Tendency of people to disperse
4. Disruption of traffic
5. Explosions and fire
6. Hazards to humans and livestock
7. Disruption of business and industrial activities
8. Evacuation

B. Potential actions at the scene	Agency responsible
1. Establish an emergency headquarters	Municipal government/EMO
2. Establish adequate communications	Police/fire department/EMO
3. Rescue and fire fighting	Fire department/engineers/EMO
4. Determine nature and effects of the gas or liquid	Engineers/medical/industry
5. Warn adjacent areas and define area of risk	Police/engineers/utilities
6. Evacuate area	Police
7. Eliminate futher escape of gas or liquid	Engineering/industry/utilities

8. Notify hospitals of casualties including number and type	Medical/police/EMO
9. Establish temporary morgue if required	Coroner/police
10. Establish a news release system including instructions to the public	Operations control group/radio stations
11. Establish emergency welfare service	Welfare
12. Establish traffic control	Police
13. Establish crowd control	Police
14. Establish evacuation routes	Police/EMO
15. Set up an inquiry service	Operations control group

C. Equipment	Source
1. Ambulances	Emergency ambulance service
2. Firefighting and rescue equipment including respirators and resuscitators	Fire/police/engineers/EMO
3. Communication equipment	Police/fire/EMO
4. Decontaminating equipment	Industry/fire/EMO
5. Mobile public address equipment	Fire department/police/EMO
6. Barricades	Engineering
7. Anti-gas clothing if necessary	Police/fire/industry
8. Emergency feeding facilities	Welfare/voluntary organizations

Source: Ottawa-Carleton Emergency Measures Organization (1970).

Disaster Scenarios

Although it is essential to know which disasters are most likely to befall a community, this knowledge alone is an insufficient base upon which to establish a disaster plan. Far more detailed information about the probable locations and scale of expected damage and casualties are required if the demands likely to be placed on personnel and equipment are to be predicted and accommodated.

Such information is most commonly developed by the use of techniques previously described. These include the application of microzonation, mathematical modeling of disaster agent impact, scenarios, gaming, and the use of field exercises. These techniques allow anticipation of the number and magnitude of agent-generated and response-generated needs accompanying specific disasters. Disaster planners are beginning to recognize the value of such detailed predictive models. For example, the Victoria *Emergency Program Guide* (1978) includes details of earthquake damage simulations compiled by Foster and Carey (1976).

PLATE 12. The movement of supplies by helicopter during extensive flooding at Wilkes-Barre, Pennsylvania. Air support is often essential in meeting disaster-generated needs (American Red Cross photograph by Carland).

Similarly, the Chatham-Savannah Defense Council (1971), Georgia, has prepared a *Disaster Control Directive* which contains a detailed description of the effects of a potential incident at the nuclear power plant at Savannah River.

Each disaster plan should include a section devoted to simulations of probable disasters, which includes predictions of the type and extent of damage, the nature and scale of probable casualties, and the various needs and responsibilities that these circumstances generate. Such figures should then be used to ensure that the steps outlined in the remainder of the disaster plan are adequate to meet likely disaster-generated requirements. Table 7.3 might be used as a matrix to aid in the consideration of each disaster type.

Relationships with Other Levels of Government

When disaster strikes, a well-prepared plan should ensure the maximum utilization of existing organizational structures, supervisory expertise, and technical skills. It should not be a time of interorganizational competi-

Table 7.3. Defining and Coping with Disaster-Related Needs

Agent demands	What is the demand?	Who will assume responsibility for the demand?	How is the demand to be met?
Warning			
Preimpact preparations			
Search and rescue			
Care of injured and dead			
Welfare			
Restoration of community services			
Protection against continuing threat			
Community order			
Response demands			
Communication			
Continuing Assessment			
Mobilization			
Coordination			
Control and authority			

Source: After Dynes, Quarantelli, and Kreps (1972).

tion, duplication, or disagreement. To prevent such misunderstandings, the safety plan coordinator can use the preparation of the disaster plan to develop formal and informal relationships between all levels of government and the private organizations and institutions that will be involved in responding to disaster-related needs. The design of the plan itself may play a useful role in the development of the required cooperation. Various levels of government and professional organizations may be invited to take part to ensure that all aspects of disaster response are adequately covered. The safety plan coordinator may encourage such groups to appoint representatives to the safety committee and to take part in later field testing of preparedness and evaluation of personnel performance. In this way relative roles can be established by mutual agreement and duplication prevented by clear definition of responsibilities.

Parr (1969) evaluated ten communities in disaster situations, noting

PLATE 13. Private agencies often play a valuable role in supplying services during disasters. Disaster plans should accomodate this anticipated assistance (The Salvation Army photograph).

that "the immediate problem in a disaster situation is neither uncontrolled behaviour such as looting nor intense emotional reaction such as panic, but deficiencies of interorganizational coordination." Parr found that probably the best way to avoid these deficiences was the rapid activation of an emergency operations center which included at least one representative from each key disaster-related organization. At a minimum, every such agency was obligated to send representatives to daily communitywide meetings at the emergency center, where each presented a report on their agency's respective activities. As Parr pointed out:

> Through these reports each organization becomes aware of what other organizations are accomplishing, and consequently ommission or needless duplication of crucial tasks is minimized. Problems arising during daily operations are considered at the meetings and generally through discussion a consensus is reached resolving the problem. Plans of action are frequently discussed and formulated.

In Cincinnati during the 1964 flood, and at Topeka after the 1966 tornado caused devastation, emergency operation centers quickly became operative and there was effective overall coordination and direction of organizational activities throughout these crises. Parr points out that this was not the case during the 1969 Glendora flood and mudslide or after the Jonesboro tornado. As a result, in the early stages of these disasters there was organizational atomization of the emergency response. In summary, no disaster plan is likely to be effective unless interorganization coordination is assured by frequent predisaster meetings and discussions and by the use of some type of emergency operations center where agencies can interact during an emergency.

Authority and Organization

Once a community has been struck by disaster the normal authority procedures may be disrupted. When this takes place and overall control is lacking, disputes concerning jurisdictions and responsibilities are fairly common. As Parr (1969) has pointed out:

> Disaster plans sometimes fail to designate a legitimate source of overall control of emergency activities, and they thus contribute to a community authority vacuum or ambiguity concerning which official, agency, or organization has the authority to make crucial decisions during a crisis.

This was not the case during the 1969 Sioux Falls floods where the community disaster plan indicated four alternative sources of authority: (1) mayor, (2) civil defense coordinator, (3) flood-control project superintendent and, (4) a long-time employee of the city light department. As Parr pointed out, "with this explicit definition of the chain of command, there was no ambiguity or lack of authority at any time during the flood crisis." In contrast, the disaster plan for Indianapolis specified that civil defense was to assume overall control after a state of emergency had been declared. It did not make clear, however, where authority lay in a situation, like the 1963 Coliseum explosion, where no such declaration was made.

Every disaster plan should also define under what circumstances and how senior levels of government and agencies, such as the Red Cross, should be called upon for assistance. In addition, the role that military personnel might play and how their cooperation is to be obtained should be carefully established.

List of All Relevant Officials, Their Telephone Numbers, and Instructions

Every disaster plan should include a detailed listing of all emergency-related personnel, their addresses, and business and private telephone numbers. In addition, the names, addresses, and telephone numbers of alternates, capable of serving in their place should they be incapacitated or unavailable, must also be included. Typically, this information is used to develop a telephone fan-out system; each individual being obligated to call two or more others in an emergency, so rapidly alerting the entire network. Such fan-outs should be designed with some overlap so that, should one link fail, the entire section dependent upon it does not remain uninformed. Care must be taken to keep all names, addresses, and telephone numbers current. Developments in computer telephone technology will soon allow such a fan-out system to be operated automatically using recorded emergency messages.

Since a large part of the disorganization that sometimes follows the impact of a disaster agent often stems from a collapse of normal communications channels, the plan should also include a directive describing what action should be taken by such officials if normal telephone contact becomes impossible. Many plans specify a predetermined meeting point, such as the emergency control center or the initiation of a courier service under such circumstances. Both alternatives have disadvantages. The former may lead to a congregation of senior officials in an area where many of them may not be most needed. This may be especially so if the specified meeting point has been destroyed or is seriously at risk. Messenger systems are slow and may put the couriers into grave danger. Other means of communication such as CB radios might also be considered.

In addition to provisions for the alerting of senior officials, plans often contain the telephone numbers of a variety of disaster-related services. The *Natural Emergency Plan* of the Ottawa-Carleton Emergency Measures Organization (1970) includes those of the Canadian Emergency Measures Organization, 13 hospitals, the Red Cross, St. John Ambulance, Salvation Army, Ottawa Valley Amateurs Radio Club, Ottawa Emergency Rescue Squadron, Canadian Forces bases in the area, the Federal Health and Welfare Radiation Protection Division, and the Weather Bureau. Also listed are the civil vehicle pool and various police, fire, and volunteer fire departments. Similarly, the *Peacetime Emergency Plan* for the Municipality of the Township of Pittsburgh (1973) includes the telephone numbers of diving services, social service agencies such as the Canadian National Institute for the Blind, and the Canadian Ski Patrol system. Obviously every municipality should pay careful attention to the wide range of official and unofficial expertise available within its jurisdiction and take every precaution to ensure that it would be available, as quickly as possible, should it be required in an emergency. Care should be taken not to rely on telephone contact alone.

Operation of Warning Systems

Effective planning seeks to meet the novel demands of disaster at the minimum economic and social cost. To achieve this goal it must focus upon both anticipated community needs and the most appropriate means of responding to them. Two types of unusual needs normally accompany disaster: agent-generated demands, caused directly by the hazard itself, and response-generated demands that occur because of the necessity to react to the damaged social fabric (Dynes, Quarantelli, and Kreps, 1972).

Perhaps one of the most obvious agent-generated demands is the need for warning. In some cases, for example explosions or earthquakes, there is usually little available time in which to issue warnings. Most disaster agents, however, either increase in magnitude or proximity and can be detected prior to impact. As a result, disaster plans must permit an optimum preimpact response so that potential life and property losses can be reduced. The problems associated with issuing warnings is described in detail earlier in this volume; suffice it to say here that warning networks and disaster plans must be coordinated so that there is an effective, rapid, and logical organized response to warning.

Preimpact Preparations

The length of forewarning, the time between detection and impact, varies greatly from hazard to hazard. Many disaster agents, however, may be preceded by a preimpact period of days, weeks, or even years. During this phase preparations for impact should be made. A well-designed disaster plan must include the necessary procedures for readying human and technical resources, and the implementation of measures to mitigate the actual impact of the agent and to limit its adverse consequences to buildings and their occupants. Such steps might include canceling the leave of trained personnel, accelerated maintenance of emergency equipment, and stockpiling of supplies. Impact might be reduced by such measures as sandbagging, drawing down oil tanks and reservoirs, barricading windows, and immunization. Consequences might be mitigated by the wholesale evacuation of people and belongings from threatened areas during the preimpact period.

Emergency Evacuation

Where warning time permits, evacuation is often the most successful response to threat. The movement of large numbers of people from an area at risk, with the minimum of stress, involves careful preplanning. Several factors must be borne in mind. Since hazards differ in their physical characteristics and scale of impact, so too must evacuation procedures. Before any evacuation is begun, care must be taken to ensure that there is, in fact, time to complete it, since residents may be more vulnerable if impact occurs during their evacuation. It is therefore essential to know how long an evacuation is likely to take and whether it can be accomplished in the time available before impact. This problem may be particularly acute if many members of the community are unable to leave unassisted. A disas-

PLATE 14. Filling sandbags in an attempt to prevent flooding. This is a typical preimpact preparation (Regina Emergency Measures Organization photograph).

ter plan may require detailed and individual evacuation plans for certain facilities such as hospitals, schools, and prisons.

The choice of routes is crucial. It is unrealistic to assume that any one road or system of roads or railways will be an optimum evacuation channel for every hazard. Ideally, microzonations should be undertaken for each potential hazard to determine the spatial variation in associated risk. Evacuation routes should then be chosen for each threat, so that the population is channeled to areas of progressively lower risk. With some hazards such as flooding, this may involve an attempt to rapidly gain in altitude. For others, such as chemical leaks, prevailing winds may be the major determinant behind choice. Evacuation routes must be easily accessible from the high risk locations. They should also be linked to alternative transportation corridors, to be used should primary routes be blocked, and free of potential bottlenecks such as bridges or tunnels that may be easily damaged. Disaster plans should include maps showing all such clearly designated evacuation routes. There is merit in including these in the telephone directory to increase public awareness. Estimates should be made of the time required to evacuate the threatened population. Obviously, this should be less than that elapsing between the issuing of a warning and the impact of the disaster agent. Where this is not likely to be the case, attempted evacuation may not be a viable safety strategy.

Naturally it is not enough simply to evacuate an area. Alternative facilities must be available elsewhere. While many evacuees are likely to stay with friends or relatives outside the threatened area, some may not be able to do so. Shelter accommodation providing meals and other facilities must therefore be available in the receiving area. In some cases, as in Darwin, Australia, total evacuation may be necessary after disaster strikes (Haas, Cochrane, and Eddy, 1976).

Shelters

In the planning of shelters for evacuees, several factors must be given careful consideration. Obviously location is of paramount importance. Shelters must be sited in low risk areas yet be easily accessible from threatened regions. To determine suitable locations, macro- and micro-scale vulnerability analysis must be carried out to determine the scale of impact of a variety of potential disaster agents. Such analysis will often indicate that different shelter locations are optimal for each hazard or group of hazards. To accommodate this situation, a variety of buildings might be designated as shelters to be used in specific disasters. Some of these may be within the boundaries of the municipality or region involved in compiling the disaster plan. Some may be required outside its jurisdiction and are best established by reciprocal disaster agreements with other municipalities. Such buildings must be evaluated from the point of view of their structural integrity in the face of various threats, such as fires or earthquakes, as well as the facilities they offer in coping with the demands of evacuees.

Ideally, such shelters should be capable of accommodating evacuees on the scale that they can reasonably be expected and should be staffed with personnel trained in providing social services. Buildings designated as shelters should be capable of storing food, clothing, and medical provisions for long periods and be equipped with communication facilities suitable for collecting and transmitting personal statistics about evacuees. Provision should be made to inform relatives and friends of the safety of the individuals concerned. Any disaster plan should also include a section dealing with the care of special evacuees from prisons, retirement homes, and other institutions that will require unusual services.

Disaster Control Center and Subcenters

Many of the problems associated with disaster stem from a lack of coordination among the varied groups and institutions that attempt to respond. Typically each views the situation in terms of its own capabilities and perspectives. Many of these difficulties can be avoided if predisaster planning has included the establishment of an emergency operations center. This should be clearly designated and recognized as such, equipped with a wide range of materials, and staffed by personnel representing all agencies involved in meeting disaster-related needs. This center should also be the site from which overall control of emergency activities is undertaken.

It is essential that, because of its key role in response, such a building is structurally very sound, sited in a minimum risk location, and unlikely to become isolated. It should not be chosen simply because of its convenience for city personnel or the low cost of purchase or equipping it. An alternative emergency command center should also be designated and made capable of functioning as such should the primary building cease to be operational. It is useful for a disaster plan to include floor plans of the major buildings described within it, including the emergency operations center. These should provide details of the storage of all emergency and communications equipment. Street plans showing the most suitable routes to such buildings should also be provided.

During the 1964 Cincinnati flood and after the 1966 Topeka tornado, emergency centers became functional very quickly as a result of preplanning. These allowed overall coordination and direction of organizational efforts throughout these crises. In contrast, such centers did not exist at the beginning of the disaster situations in Anchorage in 1964 or Minot in 1969. As a result of this lack of adequate preplanning, initial reactions to both the Alaskan earthquake and North Dakota flood were uncoordinated (Parr, 1969).

Disaster plans should also allow for the establishment of command posts at the disaster scene. Naturally their staff will be ultimately responsible to the individual with paramount authority, who will normally be at the emergency operations center. Such command posts provide members of organizations involved in the immediate on-site disaster response to receive advice and direction and allow instructions and information to be channelled more effectively. In many cases, as during the 1967 Fairbanks, Alaska, flood or the 1968 Jonesboro, Arkansas, tornado, such command posts were not established and control was largely lost. In contrast, a command post was set up after the 1965 Wichita, Kansas, plane crash, thus ensuring overall control of disaster operations (Parr, 1969).

Communications

Communication provides the basis for effective disaster response, without it other emergency-related demands cannot be met. Information is a critical aspect of every disaster; data are required about the magnitude, frequency, and location of impact, about victims, evacuees, materials, damage, the state of essential services, the capabilities and number of personnel, volunteers, and numerous other factors. Every agent-generated demand has its own related informational characteristics that must be provided when required. Information must flow during and after a disaster or coordination and effective response become impossible. There must be links between and within disaster-related organizations so that needs and orders can be communicated. They must also exist between such institutions and the public so that warnings can be issued and information on the safety or otherwise of those affected can be provided. Unfortunately, disasters often damage existing communications networks while the convergence of people and information from outside may overload those that continue to function. For these reasons, disaster planning must place heavy emphasis on maintaining existing channels and on providing additional novel alternatives such as citizen band radios, taxi telecommunications, and other potential means. Certain communication links must be preserved for emergency messages only.

Public Information

Few disaster plans make suitable arrangements for disseminating disaster-related information to community organization, the mass media, and to the public. Consequently, varying and conflicting reports of an emergency sometimes occur and receive extensive coverage. Parr (1969) notes that this was the case during the Anchorage earthquake disaster and the Fairbanks and Minot floods. To avoid this type of unfortunate development, the responsibility for making public statements should be limited to a specific individual such as the safety plan coordinator. These media releases should be carefully prepared to make sure that no erroneous information is given out or misimpressions created.

Search and Rescue

Many disaster agents result in damage to buildings and the trapping of population that requires a search and rescue response. The basic demands are for the location, rescue, and transportation of victims, often in an injured state, to places of safety where their immediate needs can be accommodated. This task often requires specialized equipment such as boats, planes, helicopters, mobile hospitals, bulldozers, and other technology together with the qualified personnel to operate and maintain them. A well-prepared disaster plan will therefore list all the sources and locations of such materials and individuals and will ensure that they are put into a state of readiness in the preimpact period, when time permits. Where obvious gaps occur in personnel or the availability of search and rescue equipment, a program of training and acquisition should be provided for in the disaster plan.

Community Order

Since confusion may follow the initial disaster-agent impact, it is essential that community order be maintained. Disaster plans should make provision for the control of the perimeter of the affected area, guarding property, patrolling danger areas, directing traffic and communications, and ensuring that scarce community resources are utilized most effectively. In all cases the ultimate authority and chain of command must be carefully established prior to the development of threat.

Medical Facilities and Morgues

The treatment of the injured and the handling of the dead are normal agent-generated demands associated with most disasters. Typically the hurt must be moved as rapidly as possible from the impact area to locations where supplies and medical personnel are capable of providing relief. Often priorities in the treatment of individuals must be established. This may involve soliciting information on case histories and determining the possible extent of injuries in the field. Such paramedical responses may often mean the difference between life and death for the more seriously injured victims. This information must be recorded in a legible manner and colored tags can be used to establish hospital treatment priorities.

Fatalities also create a disaster demand. If left unattended they themselves may generate epidemics. The dead must therefore be removed from the disaster area to some permanent or temporary morgue where they can be identified and the cause of death established and certified. Bodies can then be released to relatives and funeral arrangements finalized. These procedures require the mobilization of a wide range of individuals from coroners and dentists to fingerprint experts and morticians. Any disaster plan should therefore include details of all such hospital and morgue facilities together with information and telephone numbers of the necessary personnel.

Restoration of Community Services

The impact of a disaster-agent frequently results in the disruption of essential community services, such as electricity, gas, telephone, water, transportation, and refuse disposal. The restoration of such facilities, even if only on a temporary basis, is extremely important since search and rescue, the care of casualties, and cleanup operations are often hindered by their absence. The rapid restoration of essential community services should therefore be stressed in disaster planning. Ideally, priorities should be established and responsibilities designated so that crucial services are restored with the minimum of delay (Dynes, Quarantelli, and Kreps, 1972).

Protection against Continuing Threat

Initial disasters are often compounded by secondary hazards. These include broken gas mains and power lines, polluted waters, and fires. Any disaster plan must therefore ensure that the community is protected from such secondary, yet related threats. All potential high risk locations such as dam sites, power stations, and sewage works must be rapidly checked and damage contained. Unless such threats are quickly detected they may ultimately prove more destructive than the primary disaster agent.

Continuing Assessment of the Total Situation

Dynes, Quarantelli, and Kreps (1972) point out the significance of a further response-generated demand, the need for a continuing assessment of the emergency situation. Assessment of the actual damage caused by a hazard and of the effectiveness of responses to it are crucial. This is because without it confusion will result and organized action will be suboptimal. Assessment, therefore, is an essential prerequisite of adequate decision making. Unless a clear overview of the entire situation is available, urgent needs will go unrecognized. The significance of assessment is continuous, since the situation is likely to alter rapidly as secondary threats develop and new sources of assistance become available. Without a detailed overview, it is impossible to set priorities and to gauge the likely effectiveness of selected strategies. For this reason the task of collecting, portraying, analyzing, and assessing disaster-related information should be laid down in any disaster plan. This assignment must be realistic and should allow its accomplishment in real time, so that the ongoing assessment is available to decision makers actively responding to the disaster situation.

Reciprocal Agreements

While every effort should be made to develop self-sufficiency, reciprocal emergency agreements provide a valuable back-up mechanism. These should be negotiated in detail among a group of communities that are unlikely to suffer simultaneous disasters. Such agreements should be listed in any disaster plan and should specify the type of aid that will be provided, under what circumstances, and by whom.

Such agreements have certain less obvious advantages. They allow officials the opportunity to gain experience in actual disaster situations which do not directly affect their own community. They also provide a pool of expertise not necessarily available within the community itself. Because of their automatic nature they make sure that should a large scale emergency occur, outside assistance will be provided with a minimum of delay.

Testing the Plan

There are many aspects of disaster response that can only be tested by rehearsal. As a consequence, every plan should be subjected to appraisal through game simulations and field exercises. Following each such test and after an actual disaster occurs, the plan must be evaluated and changes made to accommodate deficiencies.

Revision

Every community is in a state of flux. New industries are established, old businesses close, and construction and demolition constantly change the pattern of land use. Materials and equipment are acquired and disposed of and individuals leave and enter the community, bringing changes to the skills and experience available in an emergency. As the fabric of a society alters, so too do both the risks it faces and its potential for responding to them.

Preferably, many of those changes should be monitored by computer, allowing rapid updating of data within the disaster plan. Whether or not this is practical the disaster response guide should be revised annually. After the 1963 Indianapolis, Indiana Coliseum explosion it was found that disaster plans had been neglected for several years, and as a result lists of available equipment and names and phone numbers were outdated. (Parr, 1969).

To make sure that this does not occur and that new information is utilized by those involved in plan implementation it is useful to publish the document in ring binders so that revised procedures can be inserted and obsolete pages discarded with the minimum of effort.

Distribution

Discussions over the distribution of disaster plans normally focus on the conflicting need for awareness and for confidentiality. Typically, the community disaster plan contains information that would be of great value to many organizations and individuals in an emergency. It also includes data, such as details of communication channels, police procedures, and expected casualties in particular disaster situations, that should not be widely distributed. This is because such confidential information may allow unauthorized individuals to interfere with emergency procedures or cause unnecessary stress within the community.

To overcome this potential conflict between the desire to know and the need to keep confidential, municipalities might consider producing both comprehensive and abbreviated versions of their disaster plan. The distribution of the former might be restricted to those involved in official response in an emergency. These individuals should be listed in an appendix to the document itself. The abbreviated version of the disaster plan, from which confidential details have been deleted, might be far more widely distributed, being made available to the media and the general public. If this approach is not considered feasible certain details, such as the significance of warning signals, evacuation routes, and shelter locations, should be published in the telephone directory and be made available in other easily accessible public places.

References

Abbott, P. L., and J. K. Victoria (Eds.). 1977. *Geologic Hazards in San Diego: Earthquakes, Landslides and Floods.* San Diego Society of Natural History, San Diego, California, 96 pp.

Chatham-Savannah Defense Council. 1971. *Disaster Control Directive.* Savannah-Chatham County, Georgia.

Chatham-Savannah Civil Defense. 1971. *Operational Survival Plan.* Savannah-Chatham County, Georgia.

Dynes, R. R., E. L. Quarantelli, and G. A. Kreps. 1972. *A Perspective on Disaster Planning.* Disaster Research Center, Department of Sociology, Ohio State University, Columbus, Ohio.

Foster, H. D., and R. Carey. 1976. The simulation of earthquake damage. *In:* H. D. Foster (Ed.), *Victoria Physical Environment and Development, 12,* Western Geographical Series. University of Victoria, Victoria, B.C.

Haas, J. E., H. C. Cochrane, and D. G. Eddy. 1976 *The Consequences of Large-Scale Evacuation Following Disaster: The Darwin, Australia Cyclone Disaster of December 25, 1974.* Natural Hazard Research Working Paper No. 27, Institute of Behavioral Science, University of Colorado, Boulder, Colorado.

Healy, R. J. 1969. *Emergency and Disaster Planning.* John Wiley, New York, 290 pp.

Mitchell, B. 1971. *A Framework for Decision Making in Water Management: A Review and Commentary.* Department of Geography, University of Waterloo, Waterloo, Ontario.

Municipality of the Township of Pittsburgh. 1973. *Peacetime Emergency Plan.*

Ottawa-Carleton Emergency Measures Organization. 1970. *Natural Emergency Plan.* Regional Municipality of Ottawa-Carleton.

Parr, A. R. 1969. A brief on disaster plans. *EMO Digest,* **9(4)**:13–15.

Santa Rosa. 1972. *Santa Rosa Emergency Service Plan.* City of Santa Rosa, Sonoma County, California.

Victoria. 1978. *Emergency Program Guide.* City of Victoria, British Columbia.

Winnipeg. 1963. *Emergency Operational Plan.* Metropolitan Corporation of Greater Winnipeg.

8
Construction and Reconstruction

Between the idea
And the reality
Between the motion
And the act
Falls the Shadow

The Hollow Men
T.S. Eliot (1888–1965)

Inevitably, safety committees will place most emphasis on preventing disaster. However, they should be fully prepared to speed community recovery if, despite all precautions, large scale damage and destruction take place. Given the resilience of modern settlements, their role within economic regions, speed of population growth, and the psychological impact of abandonment, few, if any, cities are likely to fail to recover from even major future disasters (Kates and Pijawka, 1977). The only significant center permanently destroyed by natural hazard, during the twentieth century, has been St. Pierre, Martinique. Far more typically, survivors rebuild in the same location after the wreckage has been cleared. Certain man-made hazards may make such a response more difficult. On 10 July 1976 an explosion at the Icmesa chemical plant discharged dioxin over some 4000 acres of Seveso, a suburb of Milan, Italy. Even today, 215 acres are fenced off and may never be resettled. In the vast majority of cases, however great the initial shock of devastation, decision makers should waste little time dealing with the unrealistic question of whether rebuilding should take place. Attention must instead focus on how recovery can be achieved as quickly, safely, and equitably as possible. This process of social repair is greatly facilitated if the safety committee has studied potential reconstruction prior to disaster agent impact.

Although it is unlikely that a townsite will be completely abandoned after devastation, partial relocation of certain functions is relatively commonplace. In Hilo, Hawaii, the central business district has recently been rebuilt at higher elevations to avoid further tsunami damage. Abandoned high risk sites have been converted to parkland. Where this type of partial

PLATE 15. Trailer park damage caused by Hurricane Celia, Gulf Coast of the United States. Typically, the survivors of such disasters rebuild in the same location after the wreckage has been removed (National Oceanic and Atmospheric Administration photograph).

relocation is contemplated, special effort must be taken to ensure that later expansion and subsequent population growth do not result in a return to former land use patterns. After the 1933 tsunami had completely destroyed Ryori, Shirahama, Hongo, Otanabe, and other Japanese villages, these were rebuilt at higher altitudes. Lower lying areas, once village sites, were converted to agriculture. Many unfortunately, were subsequently resettled and damaged by the 1960 tsunami, generated off the coast of Chile (Tsuruta, 1963).

The Recovery Process

It is possible to make an initial estimate of the time that recovery is likely to take after any disaster by examining the historical record. Kates and Pijawka (1977) compared the scale of population and building losses to the length of recovery time for seven devastated urban areas. Included in their survey were San Francisco, Skopje, Anchorage, Gediz, and Managua, all of which suffered extensive earthquake damage; Warsaw which was largely destroyed during the Second World War by combat; and Rapid City which was inundated by a flash flood. Their findings revealed a surprising order in recovery after disaster, indicating that total reconstruction time appears very closely related to disaster magnitude, measured in terms of loss of life and building stock.

The phase during which a community is recovering from disaster can conveniently be subdivided into four overlapping periods (Haas, Kates, and Bowden, 1977). Typically, each is approximately ten times the length of that preceding it. In the first of these, the emergency period, the disaster plan is normally in operation and attention is paid to pressing agent-generated and response-generated demands. The problems caused directly and indirectly by destruction and by the associated needs of the dead, injured, homeless, and missing provide the focus of attention. During this period normal social and economic activities are suspended or greatly modified. Depending on the nature of the disaster agent, scale of loss, and the preparedness and capacity of the society to respond to disaster, the emergency period may last for as little as a few days or as long as several weeks. It can be considered over when search and rescue operations have ceased, major transportation arteries have been cleared and are functioning once more, emergency mass feeding has been reduced, and immediate temporary housing needs have been met. The emergency period lasted for between three to four weeks in Managua, Nicaragua, following the earthquake of 23 December 1972. In contrast, it occupied only approximately one week in Anchorage, Alaska, after the 27 March 1964 seismic event (Kates and Pijawka, 1977).

The emergency period gives way to one of restoration, characterized by repairs to utilities, commercial, industrial, and residential structures. Those buildings that cannot be salvaged will be demolished and efforts will be made to return to relatively normal economic and social activities. By the end of this restoration period, major urban services, transportation, and utilities are functioning effectively once more, debris has been removed, and those refugees who intend to return to their former community will have done so. In societies with a large resource base, these activities are usually completed within a few months, elsewhere restoration may take more than a year. This period lasted for only eight weeks in Anchorage following the 1964 earthquake and for some nine months after the Managua seismic shock (Kates and Pijawka, 1977).

During the replacement reconstruction period that follows, capital stocks are rebuilt and the economy recovers to predisaster levels or higher. This phase ends when population has returned to its predisaster level and losses in jobs, residences, and urban activities have been compensated for. Certain types of reconstruction will be undertaken long after this phase is over but will be concerned with major projects. Typically, replacement reconstruction takes several years to complete, lasting, for example, for some three years in Rapid City, South Dakota after the 9 June 1972 flood.

During the replacement reconstruction period, the city generally is in a phase of flux. Dislocation inevitably follows any major disaster. Once the altruistic response to the initial impact has subsided a chain reaction of relocation, displacement, and further relocation is begun by industry and commerce and by residents capable of paying the highest rents or prices for land and property. Evolutionary patterns produced by the operation of market forces are accelerated because of the urgent need of these individuals and organizations to return to normal (Bowden, Pijawka, Roboff, Gelman, and Amaral, 1977). These forces are not restricted to damaged locations but affect a far larger part of the region. As a rule, in excess of two to three times the area is required to relocate activities from a devastated city core.

Those businesses such as banks and other financial institutions that are part of a large chain or have easy access to local capital characteristically relocate in prime locations. Similarly, long-established businesses and those with comprehensive predisaster insurance are able to meet higher land and property prices. This process was demonstrated in the rebuilding of San Francisco, where the financial institutions were among the first to relocate. In contrast, the less financially secure, smaller scale businesses are typically forced to relocate on the periphery of the commercial area. It was here in the reconstruction of San Francisco that wholesaling and industrial firms, transient residential operations, and tenements for unskilled workers and low-ranked ethnic minorities were relocated (Haas, Trainer, Bowden, and Bolin, 1977). In summary, during the replacement reconstruction period, those sectors of the economy that have traditionally dominated quickly tend to reestablish their positions of control. At the opposite end of the scale, small, undercapitalized, commercial and industrial activities may go bankrupt or leave the city completely. This process occurred in both San Francisco and Managua, leading to a decline in diversity.

The lower on the pecking order, the greater the stress, appears to be true also of socioeconomic status. In the residential sector, upper and upper-middle class individuals capable of refinancing home reconstruction or paying high rents relocate first. The lower the socioeconomic class of the individual, the more frequently market forces will create the need to make postdisaster moves and the longer the period of deferral of eventual residential stability is likely to be. In addition, such individuals will be faced with the fewest housing alternatives and be most likely to be obliged to permanently leave the city. The problem is compounded since such residents are also probably employed in low-ranking economic activities, those that are the least able to recover quickly. For this reason, those low on the socioeconomic scale may also be faced with longer periods of job dislocation and unemployment.

In the case of post-1900 disasters, a 10% loss of population has normally been replaced in three years and a 50% loss in seven years. Other factors such as financial aid and the availability of human resources and materials appear to have a less obvious impact on the speed of recovery. Indeed, Harbridge House (1972) could find no relationship between government assistance and the speed of economic recovery in major United States disasters occurring between 1960 and 1970. Despite the regularity of the relationship between disaster magnitude and rate of recovery, some cities apparently do take abnormally long or short time periods to regain their former status. A significant factor that influences their speed of recovery appears to be uncertainty, the removal of which as rapidly as possible must be a major aim of local government. It is in the provision of direction that the safety committee can play its major postdisaster role.

The simplest and most desirable way of achieving this end is through the preplanning of reconstruction. Ideally, every city should be prepared for recovery from the impact of a disaster agent. Long before devastation occurs, its planners should have produced blueprints for restoration and reconstruction. Unfortunately, this task is very rarely, if ever, accomplished, since recognition of the value of such preparation is extremely unusual. This widespread unwillingness to face up to the possibility of a major disaster increases the human suffering associated with adverse events when they do occur. It also reduces the chances of society benefiting from the opportunities for creative reconstruction that they present.

There appear to be many reasons why restoration and reconstruction plans do not sit alongside those for emergency response on the bookshelves of local government officials. Although widespread destruction occurs frequently on a global basis, individual communities almost always live with the comforting belief that they will never have to face the harsh reality of postdisaster recovery. Federal and regional governments are often confused over which of their departments, if any, should have the mandate to assist financially and technically with such disaster planning. As a result, monetary and technical aid for it is unlikely to be available.

Ideally, such postdisaster planning should be the final step in the disaster mitigation process. Plan contents should be based on estimates of death and destruction arrived at by the application of scale and mathematical models, the Delphi technique, scenario building, gaming, and field exercises. This process of compiling restoration and reconstruction plans in response to a variety of different disaster scenarios can have numerous beneficial spinoffs, even if the damage actually suffered is not that predicted. Such planning can lead to a greater awareness of deficiencies in existing policies and legislation. It will almost certainly highlight the need for new legal instruments that allow governments the rapid compulsory purchase of land in high risk areas after devastation. It will also demonstrate that mechanisms must be developed to permit expeditious changes in land use and building codes and to encourage the design of an emergency activated streamlining of the established decision-making process so that it is capable of coordinating dramatic postdisaster change in the city's infrastructure. Planning for reconstruction also pinpoints deficiencies in the existing data base, yet allows the luxury of time in which to correct those weaknesses.

Kates and Pijawka (1977) recognized a fourth phase in the recovery of a city from disaster which they termed the commemorative, betterment, and developmental reconstruction period. During this final recovery stage, building serves three differing but often interrelated functions: commemoration of the disaster, formation of future growth and development, or marking the city's postdisaster betterment or improvement. Characteristically, such projects are very large in scale and are financed by public funds. Usually their implementation occupies a time period double the length of the replacement reconstruction phase. Examples include the building of the monumental civic center complex between 1915 and 1929 in San Francisco, following the earthquake of 1906. Similarly, the development and betterment of specific locations such as the Fourth Avenue landslide area took some ten years to complete in Anchorage, Alaska, after the 1964 destruction.

Related Problems

Control

Research has shown that numerous interrelated basic issues generally arise in communities faced with the task of recovery after a major disaster. The first of these concerns the distribution and exercise of power. Several major and often conflicting goals influence the institutionalization of reconstruction. These include the desire to return to normal as quickly as possible, the need to avoid a repetition of disaster, and the belief that a better community can be built. Those who are pressing for a rapid return to normalcy, such as displaced families and interrupted businesses are likely to insist on a decision-making process that operates with the minimum of delay. This may well be some form of special task force with the power to enforce accelerated redevelopment, avoiding many of the delays due to consultation and legal restraints associated with the normal democratic process. Others may see the destruction of much of the infrastructure as an ideal opportunity to rebuild the city in a more socially equitable and safer manner. Such groups may press for orderly, well-planned reconstruction. This may lead to a demand for the exercise of power by the normal decision-making mechanisms, perhaps buttressed by outside experts and local pressure groups designed to give the public, property owners, and businessmen a special voice.

The demands of disaster are frequently so great that the local power structure is incapable of accommodating them. As a result consultants, often unfamiliar with local conditions, are hired and *de facto* control passes out of the community. Kates (1977) has argued that local officials often tend to be too involved in cleanup and search and rescue during emergency periods. These operations may best be left to specially trained personnel and that local decision makers should turn their attention as quickly as possible from the fire truck to the drawing board. This advice is sound and should be borne in mind by safety plan coordinators facing the need to respond to disaster.

Special committees should be struck to assist in the planning of reconstruction. These should include as many local experts and interests as possible so that unrealistic, grandiose, and unworkable plans are not produced by experts unfamiliar with customs, objectives, and community sensibilities. It should be emphasized that change following disaster will influence a far larger area than that which has been destroyed and a holistic viewpoint must be taken.

Another problem involved with the exercise of power is the tendency toward indecision. Within a few days of the beginning of the emergency period the reconstruction committee should have been established and a flow diagram constructed to illustrate what decisions must be taken by particular dates. Deadlines should be announced publicly and should be irrevocable. Kates (1977) has suggested that the following schedule is realistic. Within a two-week emergency period the disaster declaration to trigger land acquisition and withdrawal mechanisms should be invoked and a preliminary declaration of redevelopment, restoration, impact, and undisturbed areas made. Maps of the area involved should be published in the local press and made available in public buildings. Lease taking of land for temporary facilities should have been completed. The preliminary declaration should also allow for a 45-day moratorium in the heavily damaged or destroyed redevelopment area, permitting option taking or acquisition selection. At the end of the 45-day period the permanent four-area designation should be made.

During the time restoration is taking place, a redevelopment sketch plan should be completed with a detailed initial phase covering the first 18 months being produced. This should be subjected to public review and criticisms and revised during the early stage of its implementation. The second stage of reconstruction planning would then take place only after local opinions and insights had been provided. This would be subjected to public scrutiny and finalized before the first 18-month phase had been implemented.

This process can be speeded up if a detailed picture of the devastation is quickly developed. The structural damage in most stricken areas is surveyed between three and five times (Kates, 1977). Immediately after impact it is usual to carry out a rapid reconnaissance to provide an overview of the scale of the disaster and the type and magnitude of the required assistance. A more detailed survey often follows later during the emergency period and the process is then repeated to classify structures for demolition or reconstruction. Additional damage reviews may be carried out subsequently to aid with the redesign of building codes, the buttressing of insurance claims, or the dispensing of aid. Such a sequence of diagnostic damage surveys is perhaps unnecessarily time consuming and repetitive and may lead to indecision. This is particularly true where damage classifications are altered or buildings moved from one category to another.

Owners need to receive information on the state of their properties as quickly as possible for several reasons. Structurally sound or easily repaired buildings can be mortgaged to raise capital for reconstruction elsewhere or for relocation. They can also be rented to displaced individuals or firms, whereas those slated for demolition obviously should not be used for such purposes. Similarly, the degree of damage influences the amount and type of disaster aid for which property owners qualify. For these reasons, it is important that diagnostic damage surveys are not unnecessarily protracted. When they become so, illegal repairs may take place, including those designed to mask serious structural weaknesses. The entire process can be accelerated by careful preplanning. It should be possible to mount a rapid but careful diagnostic structure-by-structure-damage survey in the first week of the emergency period. This would be designed to make a preliminary classification of building damage similar to that carried out in Skopje, at a later date. To assist in mounting such a rapid effort, disaster plans might establish a small task force of engineers, architects, and builders which has been given the responsibility to undertake such a survey. Where disaster cooperatives, designed to provide intercity aid in an emergency, are established, assistance could be anticipated from their professional labor pool. Such a diagnostic damage task force would be expected to develop hazard-specific engineering criteria and routines long before disaster strikes. Manuals that contained drawings and photographs of hazard-related damage experienced elsewhere could be prepared in advance, as would be checklists to be completed for each building. These would provide details of the type and value of materials needed where repairs were feasible and the labor and equipment required.

Such surveys would aid in the definition of the damage zones discussed elsewhere. Copies of completed forms could be provided to owners and occupants together with guidance on repair priorities and complexity. Kates (1977) suggests that a fourfold damage classification should be used which categorizes buildings according to whether they are to be demolished, require further study before reconstruction is undertaken, can be repaired with a permit or supervision, or can be restored without official interference. He notes that the painting of the classification by color code directly onto the building has proved effective in several disaster ravaged areas.

Land Use

Although buildings may crumble or be swept away and their occupants killed or injured, the land usually survives a disaster. Exceptions occur; landslides, volcanic eruptions, coastal, and riverine erosion may actually destroy the land surface, while lava flows may greatly alter the topography. Generally, however, after the emergency period is over there remains a legacy of land ownership which greatly influences recovery. It is this pattern which tends to ensure that the new city will largely replicate the old.

Nevertheless, after a major disaster has occurred, there are usually many valid reasons for changing both land ownership and land use patterns. The impact of the disaster agent itself often demonstrates that risks have been too high and should be mitigated. In addition, there are few settlements that can be considered optimum from the point of view of efficiency and aesthetics, and destruction provides the opportunity for accelerated urban renewal, designed to improve the quality of the cityscape. In contrast, many victims are determined that the city should return to normal as quickly as possible and are unprepared for any dramatic alteration in its form or function.

It is hardly surprising that conflicts often arise over potential changes in land use. This is particularly true when large areas have been so badly damaged, for example, by floodwaters or an earthquake, that little can be salvaged. Under these circumstances the opportunity exists for major alterations in the city form. Planners often see this as an advantage, allowing well-designed urban renewal. Significant changes in land use regulations often have very widespread repercussions, well outside the area of destruction, and are the most substantive decisions to be taken following any disaster (Haas, Kates, and Bowden, 1977). Opposition to modification is most likely to come from those who will suffer financial loss from any rezoning.

One of the major errors often made by public officials during this recovery phase is the design of grandiose plans, which may take years to complete. In Anchorage, for example, an Engineering Geology Evaluation Group was established to coordinate damage assessment and delimit vulnerability zones. The American Institute of Architects recommended the formation of a committee of architects to review development plans and to ensure holistic design, while the Anchorage Planning Department drew up several alternative plans for the city based on differing assumptions about geological instability (Kates, Ericksen, Pijawka, and Bowden, 1977). Ultimately, very little of this work was incorporated into the redevelopment of Anchorage and much of the opportunity for constructive change had been lost (Schoop, 1969). This process was repeated in Managua where it took several years to complete the master redesign. By the time this plan was finished, much of the industry and commerce had already rebuilt or moved elsewhere. Advance work by the community safety committee can ensure that realistic reconstruction plans have been drawn up before any disaster strikes.

Reconstruction usually requires major changes in land use. These will either be directed by the marketplace or government fiat or, more usually, a combination of both forces. Land is required for redevelopment and also for withdrawal from the market to prevent repetition of the recently suffered disaster losses. Naturally the definition of such high risk zones should be based on microzonations. The damaged city may often require large areas of land, perhaps two to four times the heavily damaged area, for temporary residential and commercial use, for new development, and for upgrading to higher-order uses. Simultaneously, extensive areas may be withdrawn from the marketplace so that debris can be cleared and reconstruction planned. In some cases large, high risk tracts may be permanently down-zoned.

Kates, Ericksen, Pijawka, and Bowden (1977) have demonstrated that typically there are no appropriate civic mechanisms to deal adequately with these dramatic land use changes. Usually the need for land is grossly underestimated, information is inadequate, the techniques available for acquiring new land are slow and limited, funds are restricted, and existing regulations inappropriate. When the market is allowed to reallocate land after a disaster, many imperfections are apparent, such as speculation, windfall profits, hoarding, and antisocial land use.

Most cities damaged by disaster also have considerable difficulty in withdrawing land from private use. While, as this volume has demonstrated, the methodologies for identifying high risk locations are available, their application after disaster tends to be ponderous. Compensatory mechanisms and regulations governing withdrawals are circuitous and there is often great disagreement over their need and scale. As a result of this indecision, withdrawals tend to oscillate between the excessive, with high future uncertainty as in Managua, or inadequate, leading to the loss of potential for improvement, as in San Francisco.

As a result of their in-depth surveys of the reconstruction of San Francisco, Anchorage, Managua, and Rapid City, Haas, Kates, and Bowden (1977) were able to suggest a strategy for controlling land use changes in disaster stricken cities. Kates (1977) has summarized the needed procedures as follows:

> [The optimum strategy] . . . emphasizes reconstruction preparation before the disaster; the use of emergency acquisition, withdrawal and compensation mechanisms rather than normal ones; the reduction of uncertainty by rapid and simple areal designation; the giving of generous compensation for providing options where needed; and the pursuit of modest, phased and flexible planning with nonprofessional inputs, timed to the natural phases of the recovery cycle. Before the disaster strikes is the time to create an inventory and safe storage of routine information, to create the land acquisition and withdrawal mechanisms, to debate the value issues of compensation and public participation, and to designate reconstruction organization and responsibility.

Basic but essential information is often unavailable when disaster strikes, either because it has never been collected or because it has been destroyed. Every city should develop a list of available land which is suitable for temporary relocation of refugees, permanent redevelopment, and the relocation of key economic zones. This inventory should be updated regularly with information on zoning, ownership, land use, and tax appraisals. Its presence can save weeks of delay should disaster strike. In addition, total and single hazard risk maps should be prepared as part of the predisaster effort. High hazard zones should be designated by law and regulation so that their presence is an accepted tenet of the local planning conventional wisdom. In this way they can be included in postdisaster plans and withdrawals with the least ownership resistance. Kates (1977) also suggests that an annual or biennial state-of-the-city report be produced which includes details of the cash flow within the local economy, ongoing urban trends, and processes. The availability of such information cuts the time needed for study after disaster and allows a more realistic meshing of potential and reconstruction. All such basic inventories, together with cadastral maps, aerial photographs, plans of buildings, photographs, tax registers, and utility locations should be produced at least in duplicate. Copies should be stored at low risk sites which are very unlikely to be damaged by any disaster agent threatening the city itself. An exchange of such materials with another distant municipality is suggested.

Most devastated urban areas have few, if any, mechanisms for rapidly acquiring land or withdrawing it from existing ownership. Kates (1977) suggests that after impact an emergency classification should be developed. Class one land should include heavily damaged areas where redevelopment is necessary; class two should delineate that portion of the city which has been impacted but can be restored; class three areas are undamaged but are likely to be subjected to the effects of higher rents and various relocations; while class four areas are likely to remain relatively undisturbed.

In class one areas which have been badly damaged, a 45-day moratorium might be established and total acquisition of the area sought. At the end of the moratorium a plan showing desired land use should be made available and sale or options given to those willing to conform with this use. In class two areas where restoration dominates, there should be few public land purchases and emphasis placed on hazard-reducing regulation, restoration, and reconstruction. Attention should be given to removing uncertainty as soon as possible. In the undamaged area which is likely to suffer from relocation, temporary sites should be leased for emergency use. In the remainder of the urban area, that is in those designated as class four, current land use practices should continue.

Funding and compensation policies for land use changes are often inadequate and, if possible, state and federal disaster aid should be expanded to include the funding of high damage acquisitions, leases, option arrangements, and early reconstruction. Local governments might also consider the use of such instruments as bond issues, reduced future property taxes, and land swaps to produce the necessary capital to fund land acquisitions and withdrawal compensation. It is essential that generous compensation be paid as quickly as possible and that this should be at least equal to the predisaster land values.

Such dramatic changes in land use and ownership are unlikely to occur without adverse reaction. On the one hand, appeal procedures cannot be allowed to block essential reconstruction. On the other hand, individuals must not be victimized by bureaucracy without redress. Many communities have advisory groups that can act as a court of appeal. Their findings can be used to ensure citizen input in redesign.

In summary, if substantive land use regulation changes are to follow disaster, they must be made rapidly and with permanence. Any lack of authority or delays will almost certainly impair recovery. Reconstruction will be retarded, investment reduced and illegal development encouraged.

Building Codes

Disasters often present local communities with an ideal opportunity to review and strengthen their building codes, since it provides a proving ground for the theory on which they rest. Ideally, damage surveys should be carried out very rapidly to determine how structures, built to differing standards, have responded to disaster-agent impact. This process is facilitated if a file containing architectural drawings, engineerings plans, and photographs taken during and after construction have been kept on a wide variety of buildings. Timing, however, becomes most significant. It often takes months or even years to conduct detailed damage surveys and to change local building codes where necessary. Such delays stymie reconstruction and encourage illegal rebuilding. To facilitate rapid changes in codes, damage scenarios and associated new building codes should be developed in the predisaster phase by the safety committee. In this way, if specific buildings are seriously damaged by impact, and have clearly not been built to sufficiently high standards, certain changes in the local building code would be implemented virtually automatically. Such a procedure would increase safety without unnecessarily delaying reconstruction.

Aesthetics and Efficiency

Where damage has been very extensive and major changes in the form of the urban structure are possible, conflicts can arise over efforts to make sure that the city becomes more attractive or efficient (Haas, Kates, and Bowden, 1977). Such differences of opinion may become particularly abrasive if major changes in transportation routes, utility, or industrial location are contemplated. Under normal circumstances such alterations in the city fabric occur incrementally, but large scale destruction allows drastic reassessment. This may bring into conflict those who desire a return to the preexisting infrastructure with groups advocating radical change. Realistically, plans for very major change are unlikely to be implemented yet improvements can be made by compromise. As with changes in land use zoning and building codes, decisions relating to efficiency and aesthetic improvements should be taken with the minimum of delay.

Morale

Disaster-stricken communities often exhibit a high degree of solidarity, particularly during the emergency phase. There is usually a widespread and intense desire to help which does not carry with it expectations of reward or compensation. This has been called the "altruistic community" (Barton, 1969). Part of such cohesion and community spirit appears to result from the sharing of common problems and participation in collective attempts to tackle them. In the recovery stage this spirit is often less evident and conscious efforts should perhaps be made to enhance and nurture community morale. Evidence should be provided to show that residents should not be disheartened because their immediate problems are soluble and that the eventual outcome of the disaster will be community improvement.

Mileti (1975) has pointed out that the mass media, religious leaders, the mayor, governor, and other public officials can be organized to play an important role in maintaining morale.:

> The mass media provide the public with stories and anecdotes about the heroics and unselfish actions of local residents and organizations in the face of extreme odds. Local public officials make morale-boosting ap-

pearances and declare their faith in the stamina of the residents and in the community's tradition of overcoming adversity. Outside officials and organizations also support morale-boosting efforts with frequent announcements that the community does not stand alone in its hour of need, and that they can be depended upon to provide assistance and resources which may not be available locally.

This process of morale maintenance should not be left simply to chance but requires a conscious coordinating effort on the part of the safety committee. Some system of gathering information on citizens' morale and of their reactions to attempts to boost it help to facilitate this process.

Financing Recovery

Disasters involve direct financial loss by most members of the community. Such initial monetary problems are often compounded by related inflation in the price of available housing and business premises, increases in land values, and in the cost of rebuilding. Financing, therefore, is usually a major issue and pressures grow for compensation or special fiscal assistance to cover private property losses. National and regional disaster aid policies often prove cumbersome, inflexible, and their application may preclude optimum performance during restoration and reconstruction. Ideally, these should be liberalized to include the rapid funding of land acquisitions and withdrawals in destroyed areas and the cost of leases and option arrangements needed to speed reconstruction. In addition, assistance may come from private sources such as the Red Cross or from international agencies such as the United Nations Disaster Relief Organization.

As with many other postdisaster decisions, the scale and speed with which this money is distributed to victims often reflects the influence of two conflicting needs, that of returning to normal and of preventing a recurrence of disaster. If federal or state aid is tied to ensuring that rebuilding is carried out in such a way that risk is reduced, then changes in land use regulations and building codes are likely to precede substantive financial assistance. Under these circumstances the recovery of the social and business structure will be delayed. However, if compensation, low interest loans, and grants are provided without any attempt to enforce local hazard mitigation policies, a major opportunity to avoid future disaster may be lost. Preferably, this conflict can be reduced by total risk mapping in the predisaster phase together with the development of disaster-related land use and building code scenarios. In this way, much of the delay can be avoided and payments can be made rapidly without sacrificing future safety considerations.

Such potential financial assistance from international, federal, state, or provincial governments does not absolve municipalities from their obligations to prepare for disaster losses. All public buildings should be covered by available forms of hazard insurance. In this way outside aid can be channeled into the recovery of the commercial and residential sectors rather than in replacing the government's own infrastructure. The safety committee should also encourage private citizens to carry hazard insurance.

Such insurance, useful though it will be, is very unlikely to represent more than a small portion of overall losses. Far larger financial assistance is required. One method of meeting the need for substantive reserves is multicity aid agreements that include establishing a disaster contingency fund. Revenues for such an undertaking can be raised by all the cities involved through traditional channels such as bond issues, sales and property taxes, or from more exotic sources like lotteries, surtaxes on occupants of high risk zones, or tariffs levied against industries that by their presence in the urban setting increase the potential for disaster. The total accumulated capital from such a contingency fund or a prearranged portion of it would be made available to aid restoration and redevelopment in any disaster-stricken city involved in the cooperative. In addition, such agreements might include labor pools, that is, individuals who, at the expense of their normal civic employer, would be willing to be temporarily reassigned to work in a devastated municipality and aid in some aspect of its recovery.

Not only do disaster agents normally damage the physical structure of a community, they also cause widespread fragmentation of its social fabric. Families are disrupted by deaths, injuries, or the loss of home, employment, and educational, religious, and recreational opportunities. These problems in turn require many new community policy decisions which must usually be taken before major reconstruction begins. Inflation frequently becomes rampant throughout this period and authorities may have to consider imposing rent controls and the rationing of scarce commodities on the basis of need rather than ability to pay (Haas, Kates, and Bowden, 1977).

The financing of reconstruction involves increased local public expenditures when tax revenues are likely to be reduced. At the same time there are greater demands for government services, such as welfare assistance and refuse disposal. Public buildings also commonly need repair and there may be a clamor for compensation for properties that have been condemned because of damage. Similarly, the enforcement of new land use regulations, and stiffer building codes may also prove an added financial burden. In many countries there are mechanisms available to provide federal asistance for rebuilding and recovery. The safety plan coordinator should be familiar with these potential sources through communication with other communities that have recovered from similar disasters, and should see that application for such aid is made as quickly as possible.

Housing

A house provides more than shelter. Through location, form, and furnishings, a residence gives expression to the values and status of its owner. From the viewpoint of the individual disaster victim, the replacement of a destroyed residence or the rapid repair of a damaged one is usually a central issue in recovery.

Elsewhere in this volume it has been shown that there are many steps that can be taken to reduce residential damage. Total hazard mapping can be mandated and legislation enacted to ensure that new subdivisions are not built in high risk zones. Building codes must be designed to accommodate the anticipated intensity of certain disaster agents, such as earthquakes, and must be enforced with rigor. Despite such preimpact precautions, residential areas will continue to experience frequent damage. The suffering this causes can, however, be reduced by preimpact preparation.

The temporary housing that is provided for disaster victims is frequently inadequate or culturally unacceptable. In Managua, tents and Styrofoam igloos were used to house earthquake victims. These were unsuitable and were abandoned as soon as possible by their residents. Similarly, camps such as those in Malaysia housing refugees who fled Viet Nam by boat, lack sewage disposal and water supply systems. If financing is available, these problems can be avoided because there are prepackaged utilities and housing suitable for every need. Kates (1977) has suggested that a catalogue of preengineered, fabricated, and packaged utilities and buildings should be prepared. This should include specifications, use, planned capacity, availability, and prices. Ideally, such facilities should be purchased in large quantities by international agencies, national governments, and perhaps disaster cooperatives. They could be stored at airports, packaged and ready for immediate shipment. Local authorities could be notified of their availability in the preimpact stage. Since such a catalogue has not yet been prepared, it is prudent for safety committees to contact manufacturers to determine what is available and the cost of bulk purchase and rental rates. In this way there is the opportunity to compare the suitability of products and their delivery dates.

Holmes and Rahe (1967) have shown that stress created by changing residence is cumulative. For this reason, disaster victims should be subjected to as few enforced moves as possible. In seriously damaged cities, thought might be given to large scale evacuation, as was the case with Darwin (Haas, Cochrane, and Eddy, 1976). If this process is carried out to facilitate reconstruction, then evacuees should not return to anything less than semi-permanent homes. Similarly, disaster victims housed in temporary dwellings outside the devastated area should not be subjected to a series of moves before being provided with a permanent home.

In the United States, a current disaster-related policy recognizes a ranking of desirability among temporary housing types, with preference being given to the use of government-owned or financed housing. Private rentals, the minimum repair of damaged buildings, and the use of travel trailers or mobile homes in the gardens of damaged dwellings are considered less desirable but acceptable. The construction of mobile home parks for disaster victims is seen as the poorest alternative. One of the reasons for this ranking is that many specially constructed, so-called temporary housing projects will in fact remain in permanent use. Prefabricated housing in Hull, England, for example, designed to accommodate those rendered homeless by German bombing raids during the Second World War, was still in use 20 years later. Since temporary facilities often become permanent, particularly when victims are from low income families, such projects should be developed with the possibility of the later addition of facilities firmly in mind. Also it is important to attempt to allot temporary housing in a manner that will not lead to later problems because of close juxtaposition of incompatible ethnic or socioeconomic groups.

Several strategies can be implemented to reduce the need for the construction of such temporary refugee parks. Building regulations and housing codes can be relaxed in three of the suggested four postdisaster land use zones (Kates, 1977). While in the redevelopment area, building restoration and reconstruction must be prohibited completely, in the restoration area, minor repairs should be encouraged as quickly as possible under supervision. Building supplies and equipment should be provided free of charge to residents wherever possible. In the undamaged zone, immediately adjacent to the major areas of impact, the codes in force should be temporarily suspended. Encouragement should be given to erecting temporary prefabricated residences or installing mobile homes in gardens, open spaces, and elsewhere to be removed over an 18-month period as new housing becomes available. Such structures would be used to house victims at levels of rent established by local authorities. The buildings themselves might be loaned to evacuees. The size of the area covered by such relaxed building regulations and codes would, of course, depend on the number of refugees involved.

The enforced moves associated with disaster often cause considerable suffering. This can be mitigated in several ways. Supportive social services including counseling can be used to minimize unnecessary movement, and to assist with financial planning. During the emergency phase every effort should be made to salvage personal belongings since these have enormous emotional as well as financial meaning. These should be carefully catalogued and returned to their original owners as soon as possible. Victims are often widely separated from friends and relatives when housed in temporary facilities. This can increase feelings of alienation and depression, a problem that can be mitigated by a free public transportation system designed to allow such bonds to be maintained during the emergency and reconstruction periods.

Where a home has been completely destroyed, the problem of an unpaid mortgage often looms large. If the victim ceases to make payments, credit ratings can be seriously damaged and his or her ability to gain additional financing eroded. If payments are continued, many individuals are unable to purchase a replacement home. In Anchorage, 921 housing units suffered between 80 and 100% damage as a result of the 1964 earthquake (Dacy and Kunreuther, 1969). In a move without precedent, the Federal National Mortgage Association passed a ruling that forgave indebtedness on uninsured mortgages held by owners, once a token payment of $1000 had been made. Later the Alaskan Omnibus Act gave similar terms to private institutions who held mortgages on homes which had suffered damages greater than 60% of their market value. Such policies solved one problem at the expense of causing another. Unfortunately they penalize the thrifty and the elderly who are most likely to own their own homes. They also deal unfairly with those who have been prudent enough to take out disaster insurance, making it less likely that others will do so elsewhere. Government purchase of all seriously damaged and destroyed homes at their preimpact market value would at least resolve some of these inequities. An alternative might be a payment of a fixed percentage of the value of the structure regardless of outstanding mortgages or insurance.

Employment

Flux is normal in the urban economy as established services or products become obsolete and innovation creates new commercial opportunities. Disaster dramatically accelerates this system of change with diverse repercussions for the fortunes of particular trades and industries. In some activities, such as construction, it may create a prolonged period of prosperity and high employment. For others, including some of those that have had their own facilities or those of their suppliers or customers badly damaged or destroyed, it may lead to retraction, migration, or financial collapse. Some marginally viable or undercapitalized enterprises, located outside the damaged area, may be unable to compete for land or services with more prosperous displaced activities and will be obliged to relocate or will be forced out of business completely.

The key to managing this system of accelerated change appears to be the recognition that many firms, beyond those suffering from the direct impact of the disaster agent, will be adversely affected. Such secondary disaster-related losses, termed indirect costs, may be enormous. Cochrane (1974) has predicted that they will probably amount to $6 billion in the anticipated San Francisco earthquake. For this reason, disaster aid designed to safeguard employment should also be made available to firms suffering from the widespread secondary impacts of disaster. Grants, loans, and subsidies should be provided to ensure that otherwise viable firms are not lost to the region during recovery. Local government, for example, might provide new land for industrial parks where firms could be encouraged to relocate because of publicly built low rental commercial accommodation or reduced property and business taxes. Many countries have designated development regions that attract firms with various incentives. Such policies are usually designed to mitigate long-standing regional disparities of income and employment, but development zones might be temporarily widened to include normally prosperous areas struck by disaster, where these would otherwise be outside such subsidized regions. Economic counseling centers might also be established to provide small businesses with information and assistance in gaining interim financing.

Where it is clearly uneconomic to salvage certain enterprises, employees should be provided with generous and extended unemployment benefits and with retraining and public employment programs. Risk capital and expertise should also be made available to entrepreneurs so that new and more viable enterprises can be established. Ideally, to ensure that as few jobs as possible are permanently lost during the recovery phase, an ongoing survey should be conducted to pinpoint those industries and trades that are in the greatest difficulty so that assistance can be provided with the minimum of delay. Thought should also be given by federal, regional, and local agencies to preferential purchasing of supplies from companies that have been adversely affected by disaster. In this way traditional markets lost during the emergency can be partially compensated for.

In Conclusion

Most communities will never have to face the challenge of recovery after large scale destruction. Nevertheless, even when deaths and injuries from natural or man-made hazards fail to reach the magnitude of those experienced in Skopje or Rapid City, reconstruction plans can be useful tools. Since they should involve a conscious commitment to reduce future risk, they can be of great value in incremental planning. Every community evolves. As it does so, it is subjected to changing patterns of risk. A well-designed reconstruction plan, backed by a hazard-conscious safety committee, can assist in ensuring that this process does not involve unnecessary risk taking. In many cases the plan's gradual implementation can decrease the probability that it will ever have to be speedily implemented. In this way disaster may be avoided and safety goals achieved. In the final analysis, security is not a right but a responsibility.

References

Barton, A. H. 1969. *Communities in Disaster.* Doubleday, New York.
Bowden, M. J., D. Pijawka, G. S. Roboff, K. J. Gelman, and D. Amaral. 1977. Reestablishing homes and jobs: Cities. *In*: J. E. Hass, R. W. Kates, and M. J.

Bowden (Eds.), *Reconstruction Following Disaster.* MIT Press, Cambridge, Massachusetts, pp. 69–145.

Cochrane, H. C. 1974. Predicting the economic impact of earthquakes. *In*: H. C. Cochrane, J. E. Haas, M. J. Bowden, and R. W. Kates, *Social Science Perspectives on the Coming San Francisco Earthquake: Economic Impact, Prediction and Reconstruction.* University of Colorado, Institute of Behavioral Science, Natural Hazards Research Working Paper No. 25, Boulder, Colorado, 81 pp.

Dacy, D. C., and H. Kunreuther. 1969. *The Economics of Natural Disaster: Implications for Federal Policy.* Free Press, New York.

Haas, J. E., H. C. Cochrane, and D. G. Eddy. 1976. *The Consequences of Large-Scale Evacuation Following Disaster: The Darwin, Australia Cyclone Disaster of December 25, 1974.* University of Colorado, Institute of Behavioral Science, Natural Hazards Research Working Paper No. 27, Boulder, Colorado.

Haas, J. E., R. W. Kates, and M. J. Bowden. 1977. Preface to J. E. Haas, R. W. Kates, and M. J. Bowden (Eds.), *Reconstruction Following Disaster.* MIT Press, Cambridge, Massachusetts, pp. xv–xxiv.

Haas, J. E., P. B. Trainer, M. J. Bowden, and R. Bolin. 1977. Reconstruction issues in perspective. *In*: J. E. Haas, R. W. Kates, and M. J. Bowden (Eds.), *Reconstruction Following Disaster.* MIT Press, Cambridge, Massachusetts, pp. 25–68.

Harbridge House. 1972. *An Inquiry into the Long Term Economic Impact of Natural Disasters in the United States.* Prepared for the Office of Technical Assistance, Economic Development Administration, U. S. Department of Commerce. Harbridge House, Boston.

Holmes, T. H., and R. H. Rahe. 1967. The social readjustment rating scale. *Journal of Psychosomatic Research,* **11**:213–218.

Kates, R. W. 1977. Major insights: A summary and recommendations. *In*: J. E. Haas, R. W. Kates, and M. J. Bowden (Eds.), *Reconstruction Following Disaster.* MIT Press, Cambridge, Massachusetts, pp. 261–293.

Kates, R. W., N. J. Ericksen, D. Pijawka, and M. J. Bowden. 1977. Alternative pasts and futures. *In*: J. E. Haas, R. W. Kates, and M. J. Bowden (Eds.), *Reconstruction Following Disaster.* MIT Press, Cambridge, Massachusetts, pp. 207–260.

Kates, R. W., and D. Pijawka. 1977. From rubble to monument: The pace of reconstruction. *In*: J. E. Haas, R. W. Kates, and M. J. Bowden (Eds.), *Reconstruction Following Disaster.* MIT Press, Cambridge, Massachusetts, pp. 1–23.

Mileti, D. S. 1975. *Disaster Relief and Rehabilitation in the United States: A Research Assessment.* PB 242976. U.S. Department of Commerce, National Technical Information Center, Springfield, Virginia.

Schoop, E. J. 1969. Development pressures after the earthquake. *In*: R. A. Olson and M. M. Wallace (Eds.), *Geologic Hazards and Public Problems.* Region Seven, Office of Emergency Preparedness, Santa Rosa, California, pp. 229–232.

Tsuruta, E. C. 1963. *Tsunami—Its Nature and Countermeasures against It.* Report prepared for the Japanese Port and Harbor Technical Research Institute, Ministry of Transport.

Index